Astronomy Study Guide

Notes to Accompany the Online Textbook

Second Edition

Robert H. Gowdy
Virginia Commonwealth University

Cover image © NASA image.

Kendall Hunt
publishing company

www.kendallhunt.com
Send all inquiries to:
4050 Westmark Drive
Dubuque, IA 52004-1840

ISBN 978-0-7575-6219-8

Printed in the United States of America
10 9 8 7 6 5 4 3 2 1

Table of Contents

Preface

This Study Guide is intended for use with the website for the Virginia Commonwealth University General Education Course in Astronomy:

$$http://www.courses.vcu.edu/PHY-rhg/astron/$$

Because the website is organized for presenting material in class and in interactive computer sessions, it is both time-consuming and expensive to print. However I have found that many students want to have a printed copy to study when they are away from their computer and to use as a starting point for the notes that they take in class. The main purpose of the Study Guide is to provide that printed copy in a convenient form.

This new edition of the Study Guide was made necessary when the website was completely redone in 2008. The basic content of the course was changed to include a much larger focus on the planets and the search for extraterrestrial life. In addition, the material was reorganized to cover science background topics in physics, chemistry, and biology, just before they are needed instead of doing them all at the beginning. As a result, the old edition of the Study Guide is no longer useful.

Some features of the website rely on animation or interaction and cannot be reproduced here. For example, the website includes a large number of multiple-choice questions that provide explanations for each answer that is selected. Most of those questions are not printed here.

The Study Guide is organized in the same way as the website. The basic unit of organization is the "module," a self-contained unit on a single topic. Modules vary in length and complexity and are numbered consecutively throughout the text. A feature unique to the Study Guide and not found on the website is a short quiz after each module. Answers to the short quizzes may be found at the end of the Study Guide.

001: Certain as the Sunrise

001.1 The first scientific observation?

Every day we see the sun rise in the East, and set in the West. The day is always followed by night and the night is always followed by day. Our ancestors would have made this observation of the natural world. We can appreciate it emotionally and discuss it logically, but the observation itself does not come from emotion or logic. Instead, it comes from our interaction with the physical world outside ourselves, from our *experience*.

001.2 Honesty and reproducibility

In science, the word "observe" means actually experiencing something for yourself. We rely on others to be truthful about their observations and to provide enough detail about the circumstances so that we could make the observations ourselves if we wanted to. When everyone who repeats an observation experiences the same thing, the observation is said to be *reproducible* and becomes part of science.

Because we rely on the observations of others, honesty is extremely important in science. The demand for reproducible observations is a relentless enforcer of honesty. When someone reports an observation that they did not make, that faked observation will probably not correspond to what others find when they attempt it. A scientist who reports too many observations that do not turn out to be reproducible is soon ignored by other scientists because they do not like wasting their time. If it is actually proven that someone has reported faked observations, the professional reputation of that individual is destroyed.

001.3 Detail and reproducibility

To reproduce an observation, you need to know *all* of the important circumstances of the observation. For example, consider the observation that the Sun rises in the East and sets in the West every 24 hours. Someone who repeats that observation in Murmansk, Russia, during midsummer will see the Sun circle the sky without ever completely setting at all. An important detail has been omitted: The observation must be made "in this part of the world." In modern terms, informed by several thousand years of science, we would say that the observation must be made at latitudes less than 66.56 degrees north of the equator. Murmansk is at almost 69 degrees north latitude and is the largest city within the Arctic Circle.

Notice that an observation that is lacking some important detail can seem to be reproducible if it has only been tested under a limited set of conditions. A primitive society with no experience of conditions in the far north (or far south) of the Earth would naturally assume that everyone sees the sun move across the sky in the same way that they do.

001.4 From observation to model

At a very young age, infants learn to go from simple observations to mental models of what they are observing. The simplest sort of model is called "object permanence" and just means that something can be hidden or out of sight and still exist.

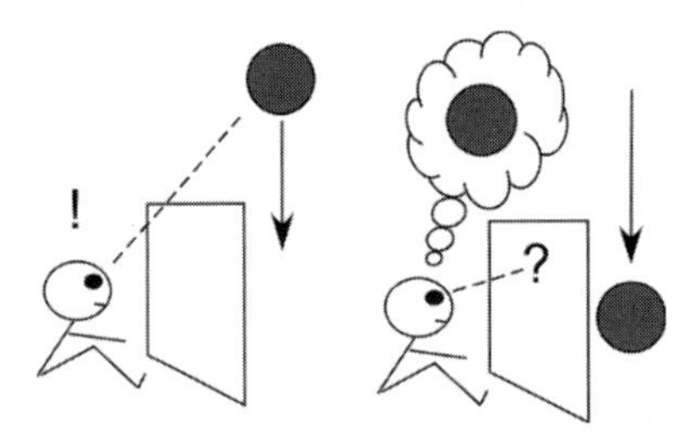

In the case of the Sun rising and setting, this model suggests that it is always the same Sun, which means that the Sun must somehow get back to the eastern horizon after setting in the west. Most primitive cultures came up with a model of the Sun's motion that looks something like this:

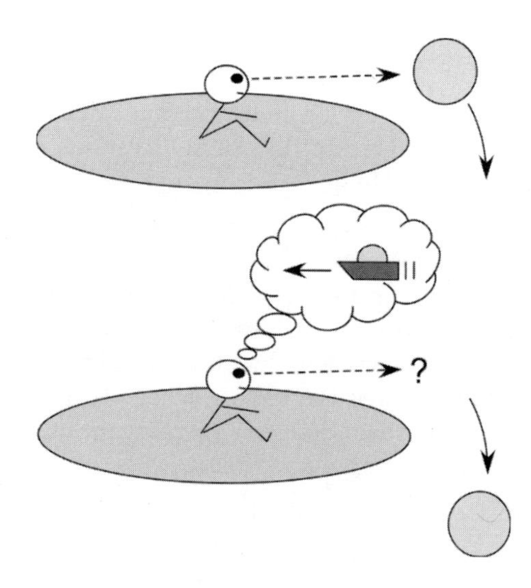

Perhaps the Sun was transported under the fat Earth in a boat.

This model provided ample opportunity for story-telling and was incorporated into ancient myths such as Apollo and his flaming chariot. However the model itself was not a myth. It made definite predictions about possible observations. Everyone on Earth should see the Sun move across the sky in the same way unless they happen to be living right on the "edge." Since the Sun spends part of its time in the "underworld" and part of its time flying through the sky, it should not be possible for anyone to see the Sun circle the sky without setting and it should not be possible for anyone to have a day without a sunrise.

001.5 Save the Appearances

The ancient Greeks are credited with the idea that this kind of prediction should be taken seriously. A worthwhile model of a situation must reproduce what is actually observed — It must "save the appearances."

The Greeks probably did not get a chance to compare notes about sunrise and sunset with travelers from north of the Arctic Circle, but they did note that travelers from Egypt reported seeing stars that could not be seen from Greece, contrary to what a flat-earth model would predict.

001 Spot Check

Here are some questions to check your understanding of the material in module 001. Both the answers and where to find these questions at the website may found at the end of the Study Guide.

1 Joseph Weber designed a series of devices to detect ripples in space-time, called gravitational waves. After several years of effort, Joe announced that he had detected gravitational waves. He built several versions of his devices and they all detected the waves. When other people tried to build similar devices, none of them detected anything at all. Joe's evidence was ignored by the scientific community because

a. Joe's observations were sloppily done and not convincing.

b. Joe was an Electrical Engineer and did not belong to the physicist club.

c. Scientists would not believe there was a possible source for gravitational waves strong enough to register on Joe's detectors.

d. Joe's observations were not reproduced.

2 The daily rising and setting of the Sun happens

a. only north of the arctic circle and south of the antarctic circle.

b. every day, everywhere on Earth.

c. only during the winter near the Earth's poles.

d. only south of the arctic circle and north of the antarctic circle.

3 Many primitive cultures have a model of the Sun's motion in which the Sun passes underneath a flat earth. This model predicts

a. that everyone on Earth sees the Sun rise and set each day.

b. that people in the far North or South might not see the Sun for many days.

c. nothing at all about how the Sun will appear.

4 Which of the following statements is an observation of the natural world?

a. I saw the Moon rise at 6:52pm yesterday.

b. There are no prime numbers that have zero for a last digit.

c. Galileo was the greatest scientist.

d. The U.S. Naval Observatory says that the Moon rose at 6:52pm yesterday.

5 One observation that we are fairly sure helped to convince the Greeks that the Earth is spherical rather than flat is

a. constellations that could be seen from Egypt but not from Greece.

b. the fact that maps of very large regions would not fit on a flat sheet of paper.

c. stories of endless nights at the North Pole.

d. that people on the far side of the Earth are upside-down.

002: Spherical Earth

002.1 Elegance

How does one come up with a mental model that can predict the results of observations? Pythagoras (born in 570 BCE) guessed that the Earth must be shaped like a sphere because that is the most elegant and harmonious shape that he could think of. It is not clear that observation played much of a role in his thinking. The flat Earth model was not rejected because it predicted things that did not agree with observation. It was rejected because it was ugly.

In modern terms, the nice thing about a spherical Earth model is that it has no edge and only one number is needed to specify the model completely, the radius R of the sphere.

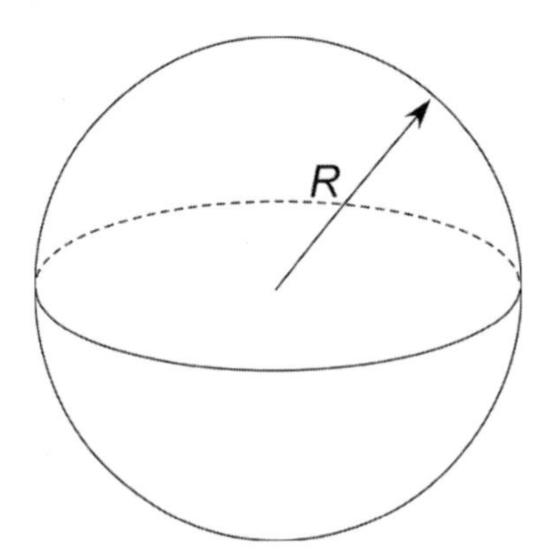

A flat Earth would have an edge of completely unknown shape and that would need far more than a single number to specify it.

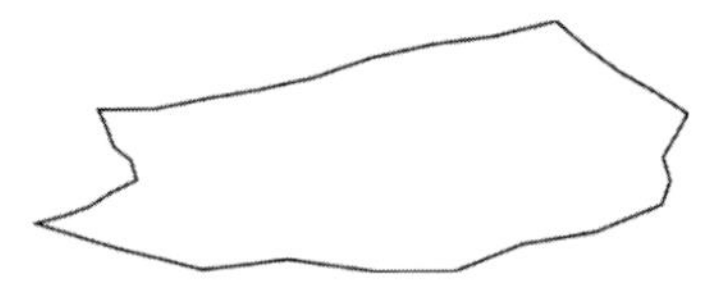

002.2 Self-consistency

Aristotle (born in 384 BCE) pointed out another attractive feature of a spherical Earth: It is *self-consistent.* If the Earth is a sphere, then gravity pulls things toward its center. However, if gravity pulls things toward the center of the Earth, that pretty much explains why the surface does not fall. It has already fallen as far as it can, which also explains the spherical shape, since all of its parts are as close to the center as they can get.

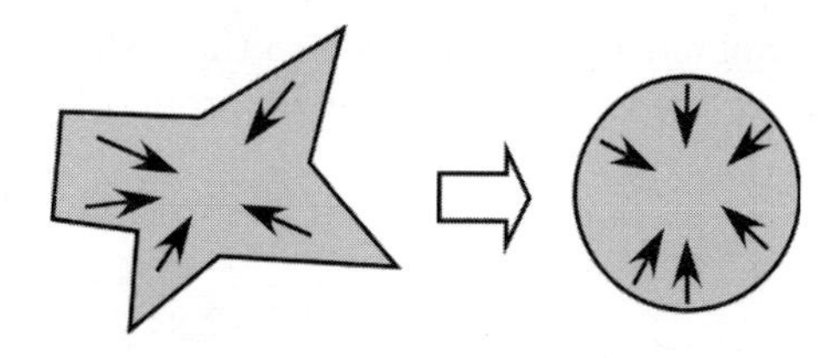

In contrast, the flat Earth fails to be self-consistent because it leads one to ask what is holding the Earth up against gravity and provides no good answer to that.

002.3 Saving the appearances

Aristotle pointed out that a spherical Earth would explain why southern constellations are seen higher in the sky in places that are farther to the south.

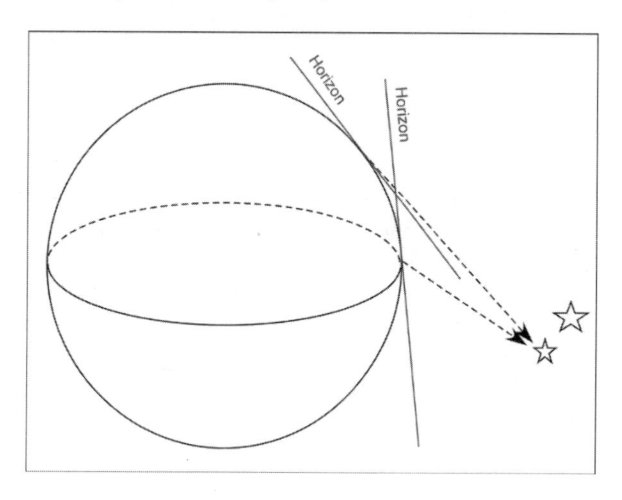

He also noted that a spherical Earth would cast a curved shadow on the Moon and that is exactly what is seen during an eclipse of the Moon.

002.4 Measuring the Earth

Eratosthenes of Cyrene knew that on the day of the summer solstice (the day when the Sun is at its highest point of the whole year), the noon Sun near the city of Syene (now Aswan, Egypt) was straight overhead. At that same time in the city of Alexandria, north of Syene, the Sun was 7 and 12/60 degrees south of overhead. The Greeks were nuts about geometry, so Eratosthenes quickly figured out what this observation implied: The angle between an Earth radius to Syene and an Earth radius to Alexandria was 7 and 12/60 degrees. That angle is just 1/50 of a full circle (360 degrees), so the distance from Alexandria to Syene must be 1/50 of the distance completely around the Earth.

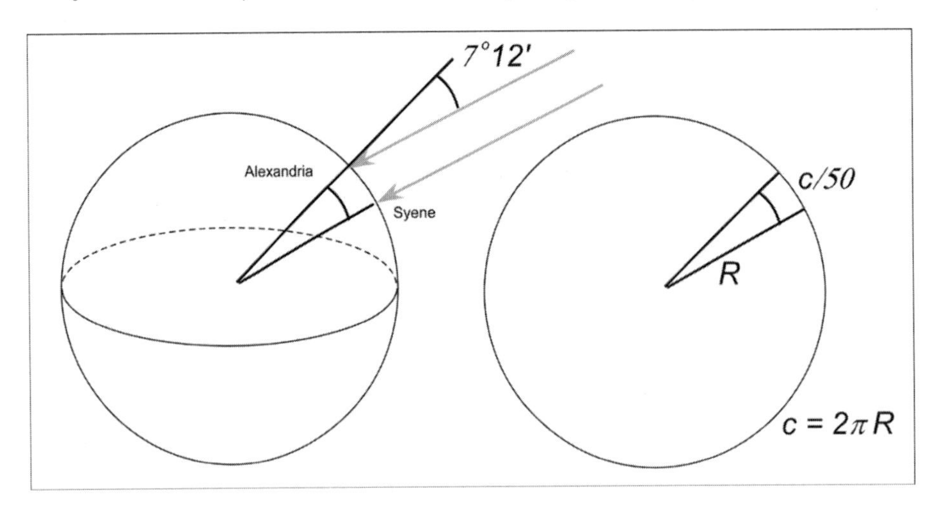

All Eratosthenes needed then was a good estimate of the distance between Alexandria and Syene.

Eratosthenes was the librarian at Alexandria and had access to extensive, fairly accurate, maps of Egypt, so he could figure out that the distance was about 5000 stadia. The circumference of the Earth had to be 50 × 5000 stadia or 250,000 stadia. A *stadion* is the length of a footrace stadium. For some reason most historians of science figure that he was using Greek stadia and the best estimate of that is 185 meters. However, since he was in Egypt and using Egyptian charts, it might have made more sense for him to use the Egyptian stadion, which was 157.5 meters.

$$c = 250000 \times 157.5\,\text{m} = 3.9375 \times 10^7\,\text{m}$$
$$c = 39,375 \text{ kilometers}$$

In that case, his value for the circumference of the Earth turns out to be really close to the currently accepted value of 40,008 for the circumference over the poles. Even with the Greek stadion, his result was not too far off.

—

A side note: The original work of Eratosthenes was lost and the details are not certain. It has been asserted that his original work filled three books and was the result of a major surveying project supported by the empire of Alexander the Great. In that case, it was probably far more detailed and painstaking than the summary given here and a precise result would not be too surprising. On the other hand, there are many historians who doubt that the project was carried out at all and was just made up by later Greek historians.

002.5 How can anyone possibly measure that?

There is no way to lay a tape measure through thousands of miles of solid and liquid rock and iron.

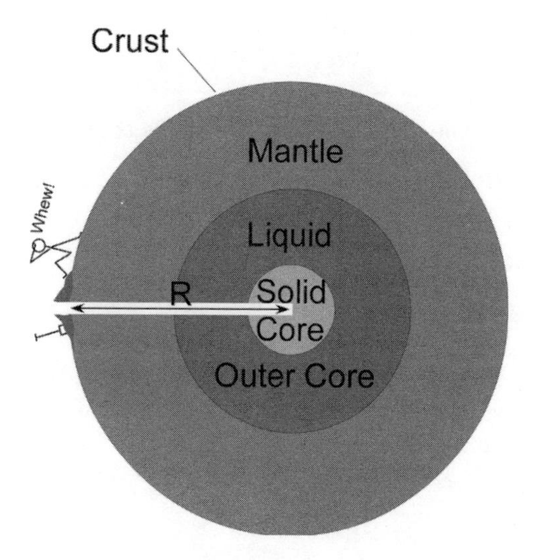

Nevertheless, by using the spherical model of the Earth and a little geometry, Eratosthenes' measurement of the circumference gives us the radius:

$$R = \frac{c}{2\pi} = \frac{39375\,\text{km}}{2\pi} = 6.2667 \times 10^6\,\text{m}$$

or

$$R = 6,267\,\text{kilometers}$$

This type of indirect measurement, using a combination of actual measurements and well-tested models, is what modern astronomy is all about. We can measure the temperature of the Sun without anyone needing to put a thermometer in it. We can measure the chemical compositions of stars without needing to travel to them. Without setting foot off of our planet, we can measure the distance to every star that we can see. We do these things with geometrical and physical *models*.

002 Spot Check

Here are some questions to check your understanding of the material in module 002. Both the answers and where to find these questions at the website may found at the end of the Study Guide.

1 One advantage of the spherical Earth model is that it completely explains

a. what holds up the surface of the Earth.

b. what causes the Sun to rise and set.

c. the retrograde motion of the planets.

d. the existence of oceans.

2 Astronomy is known for claiming to have measured many things that cannot possibly be probed directly. These measurements are made by combining actual measurements with

a. wild guesses.

b. arrogant claims.

c. fictitious measurements.

d. well-tested models.

3 The earliest known measurement of the circumference of the Earth used

a. the angle between the Sun and the quarter Moon to determine the angle between the Sun radii to the Earth and Moon.

b. the duration of a lunar eclipse to determine the angle swept out by the Moon during the eclipse.

c. noon Sun angles at two locations to determine the angle between the Earth radii to those locations.

d. noon Sun angles to determine the longitudes of two different locations.

e. the angle between the Sun and the quarter Moon at two locations to determine the angle between the Earth radii to those locations.

4 One observation that Aristotle used to justify a spherical model of the Earth was that

a. southern constellations were seen higher in the sky in Greece than in Egypt.

b. southern constellations were seen higher in the sky in Egypt than in Greece.

c. eastern constellations are seen higher in the sky in Greece than in Egypt.

d. eastern constellations are seen higher in the sky in Egypt than in Greece.

5 One property of a model such as the shape of the Earth is the number of adjustable parameters it has — the number of numbers that are needed to determine the model. If this number is very large, that is regarded as

a. of no importance so long as the model works.

b. a bad thing since it lets the model fit many possible measurements.

c. a good thing since it lets the model fit many possible measurements.

003: Celestial Sphere

003.1 A map of the stars

The stars look as if they are attached to a sphere that surrounds the earth, the Celestial Sphere. The ancient Greeks would have taken this model literally, with each star a light embedded in an actual Celestial Sphere made of crystal. We now know that stars are really at different distances from the Earth, but we still use the Celestial Sphere model as a convenient way to map the directions in which we see the stars.

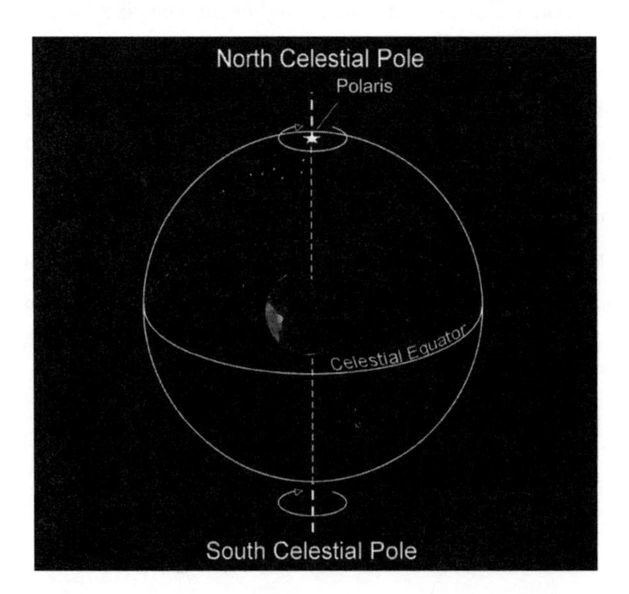

The Celestial Sphere appears to rotate from East to West, carrying each star completely around in slightly less than 24 hours. The ancient Greeks would have taken this rotation of the Celestial Sphere literally. We now know that it is the Earth rotating on its axis and moving around the Sun that makes the Celestial Sphere seem to rotate.

003.2 Constellations

Patterns in the sky

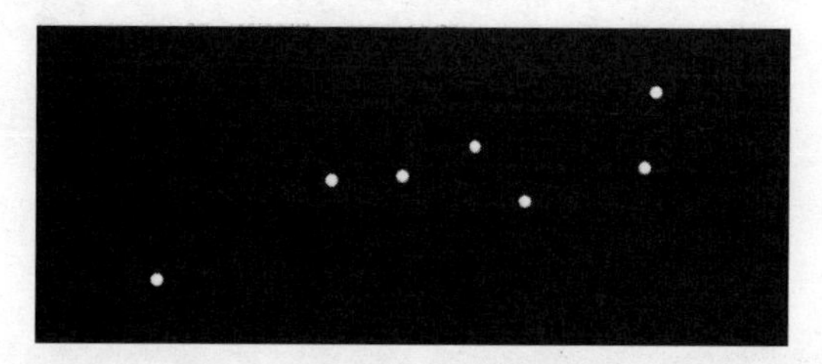

A fixed pattern of stars is called a *constellation.* The Big Dipper, shown here, is part of the constellation Ursa Major. These patterns are partly illusions because stars that look close together in the sky may be at very different distances from us. Later we will discuss how these distances can be measured, but here are the known distances for the stars of the Big Dipper:

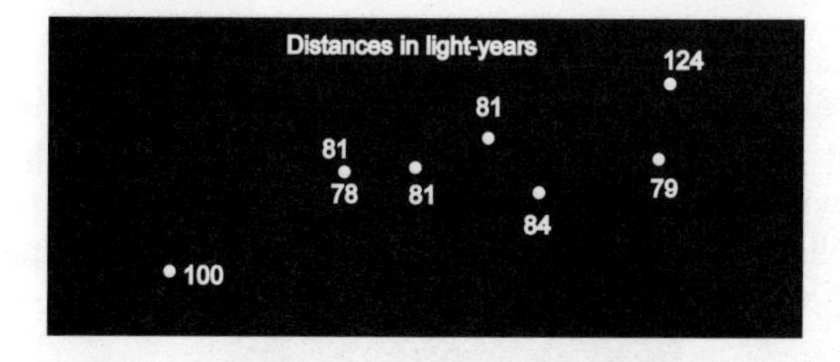

Notice that most of these stars are at almost the same distance from us, about 80 light-years away. However the two stars at the ends of the figure are at very different distances from the others and do not really belong with them.

Pointer stars

Once you have identified a constellation, you can use patterns of stars in it to locate other things. Here, the pointer stars in the Big Dipper are used to locate Polaris, a star very close to the North Celestial Pole.

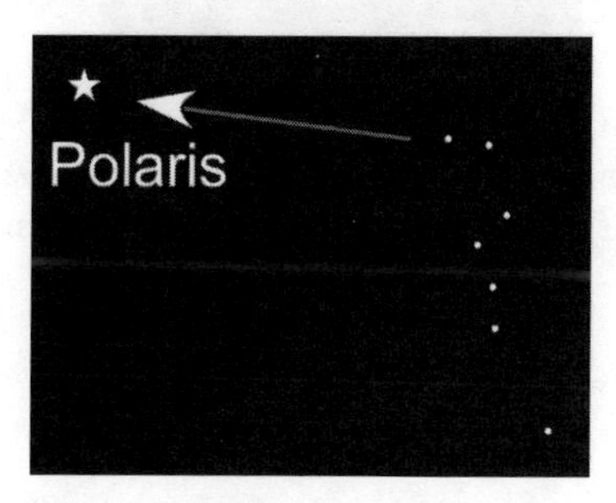

One of the most visible constellations in the sky is Orion. Here, the three stars of Orions "belt" are used to locate the star Sirius, one of the very brightest stars in the sky.

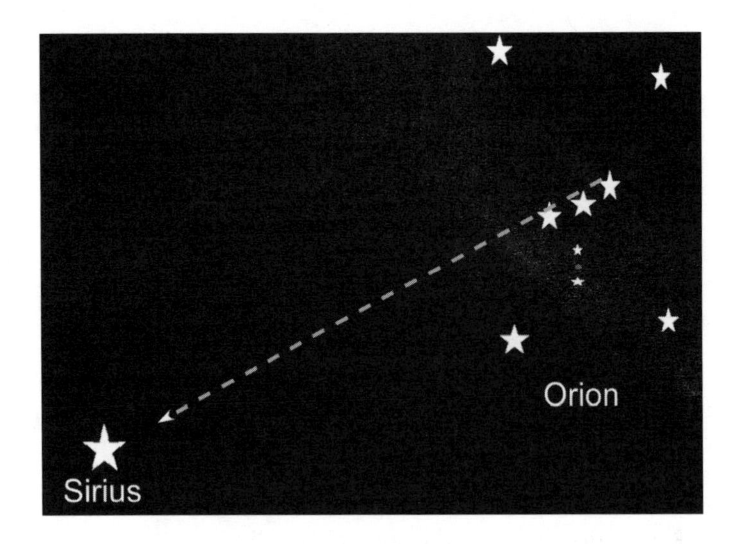

Star names

The stars within a constellation are named according to their brightness. The brightest star has a name that begins with the greek-letter alpha. The next brightest begins with the greek letter beta, and then gamma and then delta. Here is the constellation Orion, with the two brightest stars named in this way:

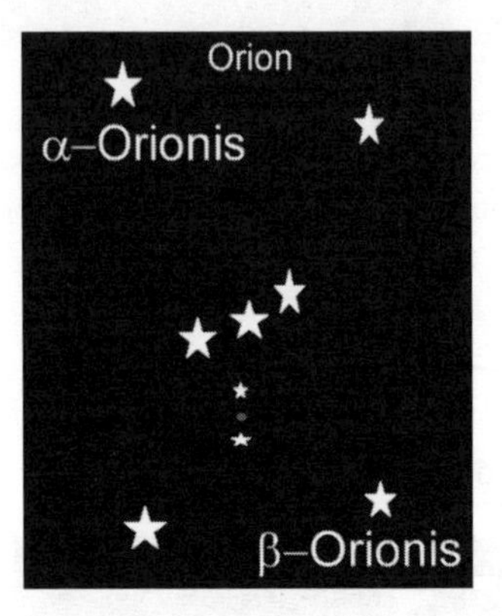

Very bright stars often have ancient traditional names. Thus, alpha-Orionis is also called "Betelgeuse" and beta-Orionis is also known as Rigel.

003.3 Apparent star motions

Some constellations, such as Orion, behave very much like the Sun, rising in the East and setting in the West.

Other constellations circle the Celestial Poles, never rising or setting. These are called *circumpolar constellations.* Here, in the middle latitudes of the Northern Hemisphere, the easiest circumpolar constellations to recognize are

Ursa Minor

Ursa Major

Cepheus

Casseopeia

You should be able to find Ursa Major, which contains the Big Dipper and the pointer stars. The pointer stars in turn, take you to the North Celestial pole, where you find the star Polaris. Polaris, which ends the "tail" of Ursa Minor (The Little Dipper), is very close to the North Celestial Pole and moves very little. The other stars circle counterclockwise around the North Celestial Pole.

The constellation Cepheus is easy to recognize because it looks like a house with the roof pointing near to Polaris. Casseopeia has a 'W' shape that is easy to recognize. The top of the 'W' is in the direction of Polaris and can be used to find it when the Big Dipper pointer stars are hidden behind buildings or trees. The remaining circumpolar constellations are more spread out and harder to recognize. Draco, for example is all over the place.

Ancient astronomers attributed these apparent star motions to the Celestial Sphere rotating around the Earth. We, of course, attribute it to the Earth rotating with respect to the distant stars. In addition to these apparent motions of rising, setting, and circling the celestial poles, stars are actually moving on their own. The stars are so distant that these *proper motions* as they are called can only be noticed over long periods of time. For example, the end stars of the Big Dipper are not moving in the same way as the rest of the stars that make up that figure. As a result, over millions of years, the shape of the Big Dipper will change.

003.4 The apparent motion of the Sun

The sun rises in the east, moves across the south, and sets in the west much as stars in non-circumpolar constellations do. Just as for the stars, this apparent motion is really due to the rotation of the earth on its axis. However, the Earth's motion around the Sun adds a complication to the Sun's apparent motion. In a full year, or 365 days, the Earth would go completely around the Sun so that the line-of-sight joining the Sun to the Earth would turn through 360 degrees. In one day, that line turns through approximately one degree.

The picture below shows the Earth rotating through a full 360 degrees relative to the distant stars. A vertical stick marks a spot that sees the Sun directly overhead when the rotation begins. During that rotation, the line-of-sight between the Sun and the Earth turns through about one degree. As a result, the Sun is not directly overhead and the Earth needs to rotate one more degree to have the Sun directly overhead at the spot where the stick is. That takes about four minutes.

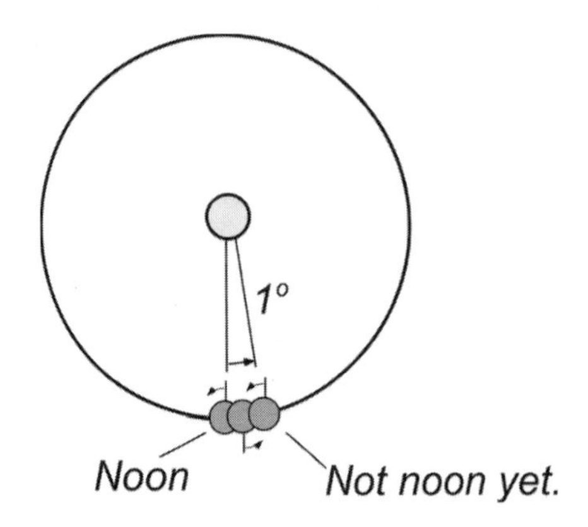

A *solar day* is defined to be the time from one solar noon to the next and is the time needed for the earth to rotate by about 361°. That amount of time is defined to be 24 hours. A *sidereal day* is the time needed for the earth to rotate by 360° and is four minutes shorter than 24 hours.

Here we have used the modern picture of the Earth going around the Sun. Ancient astronomers would have reached the exact same conclusions from their picture of the Sun going around the Earth once per year. In fact, their picture is somewhat easier to understand — Try it.

003.5 The path of the Sun

The stars that are near the sun may be seen just after sunset or just before sunrise. Observations of those stars make it possible to locate the position of the Sun relative to the Celestial Sphere.

We can say, for example, that the Sun is in the constellation Aries if we see some of the stars of Aries rising just before the Sun does and we see some other stars of Aries setting just after the Sun sets.

The path of the sun on the Celestial Sphere is a great circle, called the *ecliptic.*

The sun moves *eastward* along the ecliptic by about 1° per day, circling back to its starting point after one year.

Any object near the plane of the earth's orbit, and that includes most objects in our Solar System, will appear near the ecliptic in the sky.

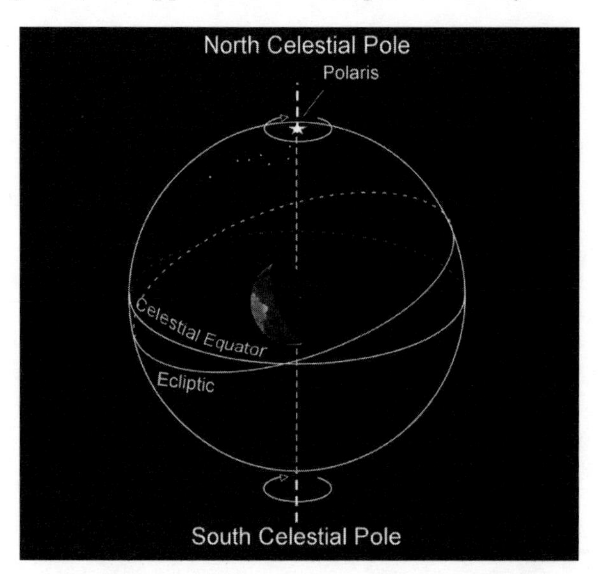

Notice that the Ecliptic great circle is tilted with respect to the Celestial Equator. The Ecliptic circle defines the plane of the Earth's orbit around the

Sun while the Celestial Equator is in the plane of the Earth's equator. The Earth's equator is tilted with respect to the plane of its orbit. As a result, the Sun rises higher above the horizon at some times of the year than it does at others.

003.6 The seasons

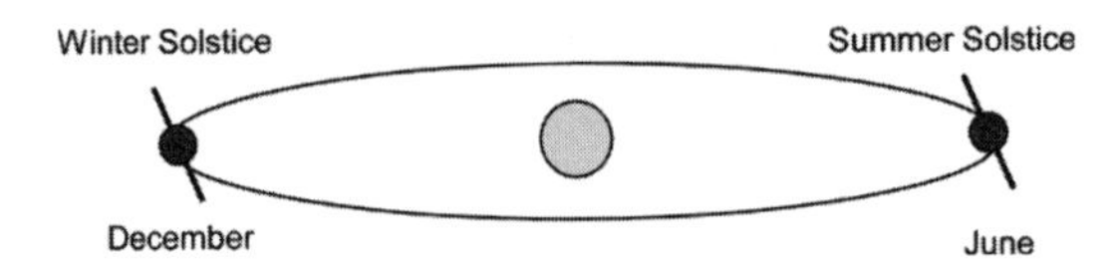

The time when the noon Sun rises the highest above the horizon is called the *Summer Solstice.* The time when the noon Sun is lowest in the sky is called the *Winter Solstice.*

Here is a better picture for seeing the effects of the season on the length of the day and on the average temperature:

Here, the Earth's rotation axis is taken to be vertical so that it is the sunlight that comes at an angle. Notice that the city of Murmansk would be near the top of the picture. As the Earth's rotation carries it around, it would stay in total darkness all day long. Obviously it would also be very cold: It would be Winter in Murmansk. A city somewhat farther south, such as Fairbanks Alaska would have the Earth's rotation carry it into the sunlight for an hour or so each day. It would still be cold. It would be Winter in Fairbanks and everywhere else in the northern hemisphere. In contrast, the folks at McMurdo Station in Antartica would be in full sunlight all day long and would be busy flying in fuel and other supplies while it is warm enough for airplanes to operate. It would be Summer in the southern hemisphere.

Six months later, this picture would have the light and dark areas exactly reversed: Long days and lots of sunlight in the northern hemisphere, short days and not much light in the southern hemisphere. The nuclear powered icebreaking ships based at Murmansk would be taking tourists through the remaining sea ice to the North Pole. Meanwhile, at McMurdo Station, aircraft would fly only in extreme emergencies because of the intense cold and unrelenting darkness.

Besides the length of the day, the angle at which sunlight strikes the Earth plays an important role in the seasons. When it is summer at a given location, the sun rises high above the horizon and the sunlight strikes the ground at a steep angle so that more heat is delivered to each unit of area. When it is winter at that location, the sun stays low in the sky and the sunlight strikes the ground at a shallow angle, delivering less heat to each unit of area.

003 Spot Check

Here are some questions to check your understanding of the material in module 003. Both the answers and where to find these questions at the website may found at the end of the Study Guide.

1 At 8pm, you see that the pointer stars of the Big dipper and the star Polaris are arranged in a vertical line. at what time would you see them arranged in a horizontal line?

a. It will never happen.

b. 7:56:00 p.m. the next day.

c. 1:59:00 am the next day.

d. 10:59:30 p.m. that same day.

e. 9:59:40 p.m. that same day.

2 An area of the sky that is marked by a recognizable pattern of stars is called

a. an asterism.

b. a stellar neighborhood.

c. a celestial sector.

d. an astral house.

e. a constellation.

3 A sidereal day is

a. several hours longer than a solar day.

b. a few minutes longer than a solar day.

c. a few minutes shorter than a solar day.

d. just the same as a solar day.

e. several hours shorter than a solar day.

4 In one day, the position of the Sun on the Celestial Sphere

a. does not change at all.

b. moves westward along the ecliptic by 1°.

c. moves westward along the ecliptic by 15°.

d. Moves eastward along the ecliptic by 15°.

e. moves eastward along the ecliptic by 1°.

5 The Summer Solstice is the time when

a. the noon sun is lowest in the sky.

b. the noon sun is highest in the sky.

c. the sun crosses the Celestial Equator.

6 The time it takes for the Celestial sphere to rotate once relative to the Earth is called

a. a sidereal day.

b. a polar day.

c. an astronomical day.

d. a Celestial day.

e. a solar day.

004: Wandering Planets

004.1 Retrograde motion

The planets appear to execute "loops" in the sky.

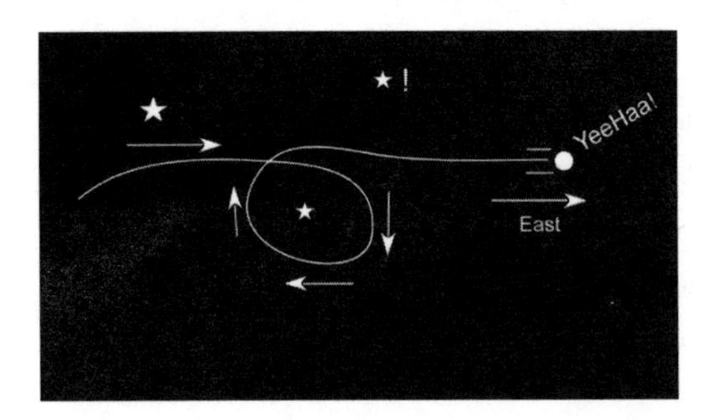

Usually they move eastward as the Sun and Moon do, but sometimes they loop back the wrong way. They show *retrograde motion.*

This behavior drove the ancient astronomers nuts!

"Obviously" the Earth itself does not move (Do you feel it moving?). To account for motions in the heavens they assumed:

- The stars are attached to a crystal sphere that rotates around the Earth.
- The Sun and Moon rotate around the Earth separately from the crystal sphere.
- Each planet moves around the Earth on a complicated path.

004.2 The Ptolemaic model of the Solar System

To model the paths of planets, the ancients assumed that each planet follows a small circular path (the *epicycle*) around a point (the *deferent*) which, in turn follows a circular path around the Earth.

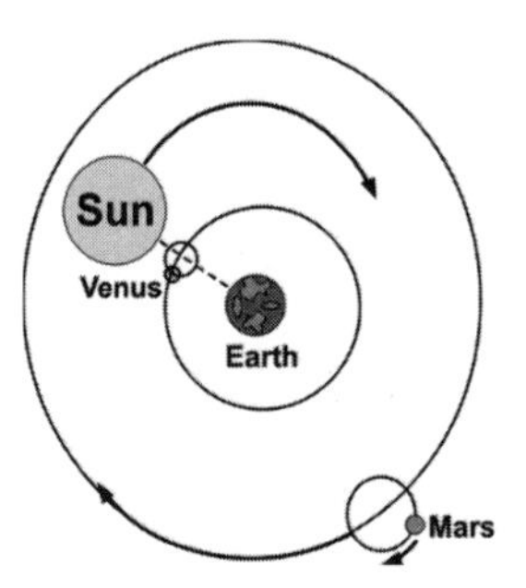

In order to reflect the actual paths of the planets, the model became more and more complex. In the final version, due to Ptolemy in about 140 AD, the epicycles and deferents were allowed to be off-center and things could speed up and slow down as they went around and there could be epicycles attached to epicycles.

004.3 Aristarchus of Samos

The ancient Greeks, starting with Aristarchus, knew perfectly well that the Sun is much larger than the Earth and Aristarchus, pointed out that it seemed pretty silly to have the giant Sun orbiting the tiny Earth. Nobody listened to him.

Aristarchus proposed that the daily motions in the heavens, the rising and setting of the Sun and Moon and the seeming rotation of the Celestial Sphere, were due to the Earth rotating on its axis. He also proposed that the Earth was circling the Sun and not the other way around. His contemporaries did not think much of his ideas. Here is a quote from one of his contemporaries, Dercyllides, which comes to us by way of the historian Plutarch:

> (Dercyllides) says that we must suppose the earth, the Hearth of the House of the Gods according to Plato, to remain fixed, and the planets with the whole embracing heaven to move, and rejects with abhorrence the view of those who have brought to rest the things which move and set in motion the things which by their nature and position are unmoved, such a supposition being contrary to the hypotheses of mathematics."

A geometrical argument against a sun-centered model is often cited: If the Earth moves around the Sun, we should see nearby stars shift back and forth relative to more distant stars. The effect is called *heliocentric parallax* and is discussed in detail later in our course. Aristarchus was well aware of the effect and explained that the stars are so far away that the shift is too small to see. He was correct. The shift actually occurs, but is so small that it was not seen until the development of powerful modern telescopes. However, his contemporaries were not willing to accept the vast size of the universe that he was proposing.

In fairness to Aristarchus' contemporaries, the Greeks had a static idea of how gravity works — Objects of the sort that the Earth is made of fall toward the center of the universe. That idea played an important role in establishing the spherical Earth as an elegant and self-consistent model and it definitely implied that the Earth would be at the center of the universe. Aristarchus was proposing that they should simply abandon that theory of gravity. In return, they would get a theory that is definitely harder to think about than the Earth-centered model and does not do a better job of 'saving the appearances.' The choice must have seemed like a no-brainer at the time: Stay with the elegant model that is easy to think about.

004.4 Doubts about the Ptolemaic System

Copernicus pointed out that the final form of the Ptolemaic system actually had the position of the Earth displaced from the centers of the various deferents in order to agree with observations. In other words, it did not really put the Earth at the center of the universe and was, itself, abandoning the Greek theory of gravity. To Copernicus, that meant that Aristarchus's idea was worth detailed consideration. If the Earth is not going to be at the center of the universe anyway, what is wrong with having it orbit the Sun?

004.5 Retrograde motion

From Aristarchus, Copernicus took the idea that the daily motions in the heavens — the rising and setting of the Sun and Moon and the turning of the Celestial Sphere — are actually due to the rotation of the Earth on its own axis. Also, like Aristarchus, he figured that the motion of the Sun along the Ecliptic and the seasons of the year are due to the Earth moving around the Sun. The new idea that Copernicus added was that most of the crazy looping motions of the planets is due to the fact that we are observing them from a platform that is itself going around in circles as it orbits the Sun.

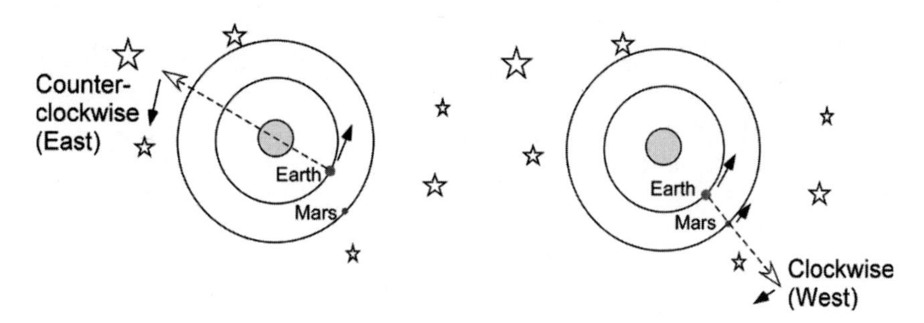

Mars, for example, appears to move westward relative the stars when the Earth passes between it and the Sun but appears to move eastward when it is on the other side of the Sun from the Earth.

004.6 Why Copernicus lost

In addition to pointing out that the retrograde motion of the planets could be explained, Copernicus also set out to produce a completely worked-out model of planetary motion. He used the same basic geometrical scheme as Ptolemy with each planet moving along an epicycle, but with the centers of the epicycles moving on circular orbits around the Sun. With the benefit of hindsight, we know that this use of circles and epicycles was a mistake. It has been asserted that Copernicus reduced the number of epicycles from 80 to 34. However, by the time (1543) that Copernicus published his final work, *De Revolutionibus Orbium Coelestium*, he needed about the same number as Ptolemy did to get comparable agreement with observations.

In the end, Copernicus was proposing a new, unfamiliar, and hard-to-understand system that was just as complex as the familiar Ptolemaic system and did not do any better at reproducing observations. The response of his contemporaries was just what you would expect: Stay with the familiar, established model. Only a few were persuaded by the aesthetic argument first made by Aristarchus, that having the gigantic Sun orbiting the much smaller Earth is just silly.

004 Spot Check

Here are some questions to check your understanding of the material in module 004. Both the answers and where to find these questions at the website may found at the end of the Study Guide.

1 Copernicus said that the Earth and planets orbiting the Sun caused

a. The rising and setting of the Moon.

b. The phases of the Moon.

c. The daily motions in the heavens.

d. The rising and setting of the Sun.

e. The retrograde motion of the planets.

2 The astronomers of Copernicus's time rejected his model of the Solar System mostly because

a. they failed to understand it.

b. it did not account for observations any better than the Ptolemaic System

c. the Ptolemaic System was supported by the Church.

3 Retrograde Motion refers to the

a. eastward motion of the planets relative to the stars.

b. westward motion of the planets rclative to the stars.

c. eastward motion of the Sun relative to the stars.

d. westward motion of the Moon relative to the horizon.

e. westward motion of the planets relative to the horizon.

4 For Copernicus, a critical inconsistency in the final version of the Ptolemaic System was that it

a. did not really have anything fixed in place at the center of the universe.

b. had the Earth fixed in place at the center of the universe.

c. had the Earth moving around the Sun.

d. had the Sun moving around the Earth.

5 One objection to a solar system model that has the Earth moving around the Sun is that we would then see nearby stars seem to shift back and forth relative to more distant stars. The correct answer to that objection is that

a. each star is moving in a circle exactly in step with the Earth's motion around the Sun.

b. gravity bends starlight in just the right way to undo the shift.

c. all of the stars are attached to the same Celestial Sphere, so none are more distant than others.

d. even the closest stars are so far away that the shift is very small.

005: The Power of Careful Observation

005.1 Big Science

Tycho Brahe was the first person to point out that whenever a comet appeared, observers all over Europe saw it in the same constellation. That contradicted the popular theory that comets were atmospheric phenomena. It meant that comets were up there moving through the supposedly unchanging heavens. He also used improved astronomical instruments to produce the most accurate star catalog of its day and made careful observations of a supernova that appeared in Casseiopeia in 1572. Just as happens today, a reputation for doing careful work and making big discoveries attracts big government grants. In Tycho's case, he became the court astronomer to King Frederick II of Denmark, was given an island and a castle to house his operation, an army of skilled craftsmen to build astronomical instruments, and a paper mill to produce the paper to write the observations on.

Tycho's castle (Uraniborg) and associated observing instruments cost about 1% of the budget of Denmark over the eight years that it took to construct them. That makes it one of the most expensive scientific projects of all time. Current Big Science projects such as the Large Hadron Collider being built in Geneva for 6 billion euros are extremely small by comparison.

005.2 Tycho's observations

Before Tycho, astronomers usually measured the positions of planets only when they were doing something interesting such as appearing near another planet or star (called a *conjunction*). In order to justify the huge expense of his observatory and keep his army of workers and research collaborators busy, Tycho attempted to measure the positions of

- all the planets
- all the time.
- more accurately than ever before

Telescopes were not yet in use, so these observations were done with large angle-measuring devices.

Before Tycho, stars and planets were usually located with an error of several degrees of arc. You can easily do as well as that with home-made instruments. (In fact, you will have a chance to do that as part of your project.) With the large and precise instruments that Tycho built, he could measure angles to 1/60 of a degree of arc — an enormous improvement.

These devices and Tycho's army of observers generated an enormous amount of accurate data on the motion of the planets. Now someone needed to figure out what it all meant.

005.3 Kepler: Putting Copernicus to the test

Johannes Kepler was Tycho's assistant near the end of Tycho's life and inherited his mountain of observed data about the motion of the planets. Since Kepler was an advocate of the Copernican system, he first used Tycho's observations to test the Copernican System.

It failed: There was no way to adjust the Copernican system to fit the accurate observations of Tycho. The Ptolemaic System did no better, but that was little consolation.

005.4 Kepler's war with Mars

When an airplane crashes, aviation safety technicians work day and night to figure out what went wrong with it. Similarly, The Copernican system had crashed and Kepler needed to figure out what was wrong with it. He started with the assumption that the basic idea of the planets going around the Sun is correct. The only thing left that could be wrong with the Copernican Theory is the actual shapes of those orbits.

Tycho's observations showed the directions in which planets were seen at frequent intervals for many years. Kepler assumed that these observations were made from a point that circles the Sun once a year. He decided to use the data to plot out the actual shapes of the orbits. The largest and most complete set of Tycho's observations were of Mars and those were also impossible to fit with any combination of circles and epicycles, so Kepler tackled those first.

In principle, it is easy to see what Kepler had to do. First determine the approximate time that Mars takes to complete one orbit by noting when Mars and the Sun were in conjunction and opposition. Those are called *synodic periods* and were already quite well known by astronomers. Refine that value by noting how long it takes for Mars to come back to a given position in the sky at a given time of the year. Those times would always be integral multiples of the time it takes Mars to complete one orbit. Then use Tycho's mass of data to find observations of Mars at times exactly one Mars period apart.

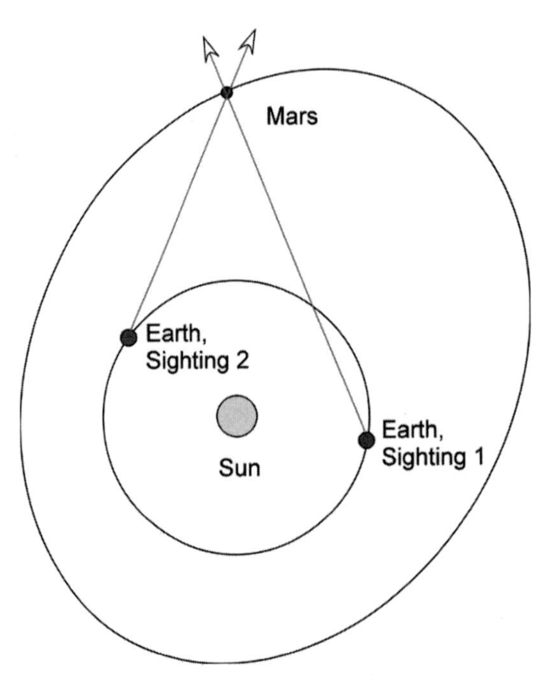

With lines of sight to Mars known from two different locations of the Earth in its orbit, the position of Mars could be located at the intersection of those lines of sight. Then repeat the process over and over until the full path of Mars was revealed. Finally, guess what sort of path that is.

In practice, without computers to do the calculations, Kepler had to be extremely clever in organizing his analysis and referred to this task as his 'War with Mars.' His progress was not straightforward at all. In 1602 he actually found the equal-area law that we will discuss later. Only after that and approximately 40 failed attempts did he decide to try elliptical orbits. He figured that ellipses were just too simple for people to have overlooked. When Kepler was done (1605), he could see that the orbit of Mars was indeed an *ellipse.* It was fortunate for Kepler that the orbit of Mars is quite far from circular. The Sun-Mars distance varies by about 9% during each orbit.

005.5 Kepler's First Law of Planetary Motion

What Kepler had found is now known as his First Law of Planetary Motion.
1. *The orbit of each planet is an ellipse with the Sun at one focus.*

An ellipse can be drawn by sticking two pins into a piece of paper and stretching a circular loop of string around both pins and the tip of a pencil. Moving the pencil so as to keep the loop stretched tight will cause it to draw an ellipse.

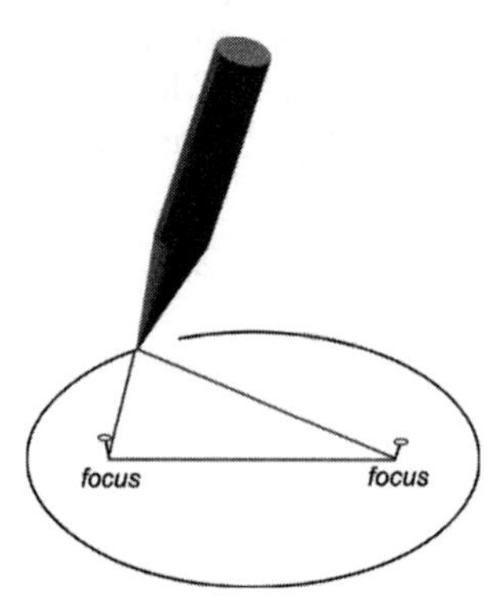

The two pins are the foci of the ellipse. An ellipse can be exactly circular when the two foci are both at the same point. When the two foci are far apart, the ellipse is long and thin. For the Earth's orbit, the two foci are extremely close together so that the orbit is nearly circular. For Mars, they are farther apart.

Here is a planet on an elliptical orbit, shown at its closest approach to the Sun, called the *perihelion* and at its farthest distance from the Sun, called the *aphelion.*

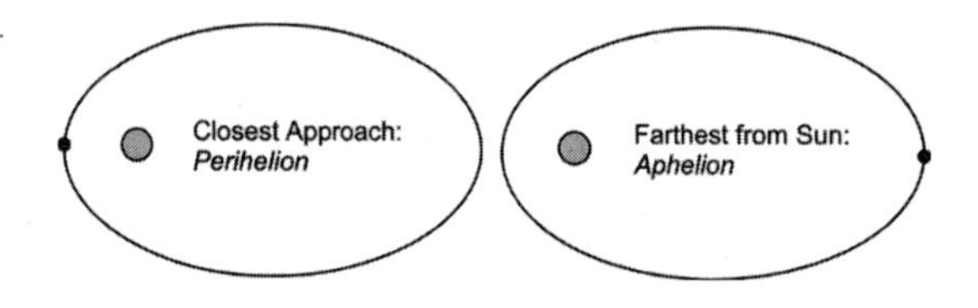

005.6 Keplers 2nd and 3rd Laws

Once Kepler figured out the actual shapes of planetary orbits, the rest fell into place. He had already found earlier that each planet speeds up and slows down according to the equal area rule:

2. *A line drawn from a planet to the sun sweeps out equal areas in equal times.*

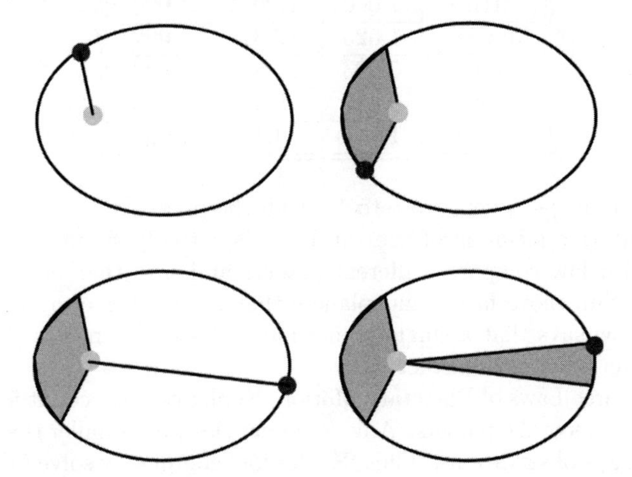

Notice that this rule, *Kepler's Second Law of Planetary Motion*, implies that planets move the fastest when they are closest to the Sun.

Finally, Kepler found that there is also a simple relationship between the motions of different planets.

3. *For each planet, let T be the time that planet takes to go once around the Sun and let R be that planets average distance from the Sun. More precisely, R is half of the longest diameter of the planets elliptical orbit. The quantity*

$$\frac{T^2}{R^3}$$

has the same value for every planet.

For example, try measuring T in Earth years and R in terms of the *astronomical unit*, which is defined to be the average distance of the Earth from the Sun. For the Earth, the calculation is easy:

$$T = 1 \quad R = 1$$

$$\frac{T^2}{R^3} = 1$$

Use modern values for the other planets, including some that Kepler did not know about:

planet	**R/au**	**T/yrs**	**T^2/R^3**
Mercury	0.387	0.240	0.994
Venus	0.723	0.615	1.001
Earth	1.000	1.000	1.000
Mars	1.523	1.881	1.002
Jupiter	5.203	11.86	0.999
Saturn	9.540	29.46	1.000
Uranus	19.18	84.10	1.002

The result is that $\frac{T^2}{R^3}$ is very close to 1 for all the planets.

The qualitative meaning of the third law is actually similar to the second law: The third law compares different planets and says that planets that are closer to the Sun move faster and planets that are farther away move slower. The second law says that a single planet moves faster when it is closer to the Sun and slower when it is farther away.

With his three Laws of Planetary Motion, Kepler could account for all of Tycho's observations of the planets. Any descrepancies were smaller than the likely errors in Tycho's observations. Thus, Kepler had completely solved the problem of how the planets move. These results were published in his *Astronomia Nova* in 1609.

005 Spot Check

Here are some questions to check your understanding of the material in module 005. Both the answers and where to find these questions at the website may found at the end of the Study Guide.

1 Tycho Brahe's careful observations of the planets agreed, to within observational error, with

a. None of these systems.

b. the Copernican System.

c. the Ptolemaic System.

d. the Tychonic System.

2 The statement that the orbit of each planet is an ellipse with the Sun at one focus is due to

a. Newton.

b. Copernicus.

c. Kepler.

d. Galileo.

3 In addition to measuring the positions of all the planets all the time, Tycho Brahe measured them to an accuracy of

a. one degree of arc.

b. 1/60 degree of arc.

c. 1/10 degree of arc.

d. 1/3600 degree of arc.

e. five degrees of arc.

4 One reason that the Copernican System failed to fit Tycho's observations was that

a. it placed the Earth at the center of the universe.

b. it used elliptical orbits instead of circles.

c. it placed the Sun at the center of the system.

d. it used circular orbits instead of ellipses.

5 When compared to planets that are closer to the sun, planets that are farther from the sun

a. move slower on the average.

b. move faster on the average.

c. move at the same average speeds.

006: Scientific Statements

006.1 Falsifiability

Statements that belong in science must be about reproducible observations. However, as Karl Popper pointed out, there is a much stricter requirement.

A scientific statement is one that could possibly be proven wrong.

Such a statement is said to be *falsifiable.* Notice that a falsifiable statement is not automatically wrong. However a falsifiable statement always remains tentative and open to the possibility that it is wrong. When a falsifiable statement turns out to be a mistake, we have a way to detect that mistake and correct it.

You might suppose that any mistaken statement can eventually be found out and corrected. However, when you are asked to say, in concrete terms, what reproducible observations could possibly convince everyone that a given statement is wrong, you often find that there are none.

Examples

Consider the following statements:

An alien spaceship crashed in Roswell New Mexico.

A giant white gorilla lives in the Himalayan mountains.

Loch Ness contains a giant reptile.

In each case, if the statement happens to be wrong, all you will ever find is an absence of evidence — No spaceship parts. No gorilla tracks in the Himalayas. Nothing but small fish in the Loch. That would not convince true believers in those statements. They would say — "The government hid all of the spaceship parts." "The gorillas avoided you and the snow covered their tracks." "Nessie was hiding in the mud at the bottom of the Loch." In other words, *absence of evidence is not evidence of absence.* None of these statements is falsifiable, so none of them belong in science.

Notice that even imagining going to silly extremes is not enough to falsify these statements. What if you drained the Loch and put all of the mud at the bottom through a strainer and found nothing? The Nessie believers could just say that Nessie got away through a hidden passage to the ocean before you could get all the water out. The point is that we live in a universe with an endless ability to surprise us, so nothing can ever be proven or disproven by exhausting all of the possibilities. There will always be possibilities that we have not thought of.

What about the opposite kinds of statement:

No alien spaceships have ever landed in Roswell New Mexico.

All you have to do here is find just one spaceship and the statement is disproven. An exhaustive elimination of possibilities is not needed. Just one spaceship will do it.

Do you have the idea that only negative statements can be falsifiable? Consider this one:

This critter (just pulled from Loch Ness) is a fish.

Just one observation — "Uh, it has fur all over it." — is enough to disprove this statement, so it is falsifiable. Biologists love classification statements like this one because they are always falsifiable.

006.2 Speculations

Statements that are not falsifiable cannot be disproven if they happen to be wrong. The confusing thing is that often these same statements can be proven if they happen to be right. For example, the statement that Loch Ness contains a giant reptile could certainly be proven by snagging a giant reptile and hauling it up onto the boat dock. This type of statement, provable if it happens to be right, but not falsifiable if it is wrong, is called a *speculation*.

One of the objections to Popper's philosophy of science is that real scientists are often driven by speculations and sometimes they turn out to be right. A biologist, for example, might be driven by the idea that the ivory-billed woodpecker is not extinct. That is a speculation because it is not falsifiable. If the speculation should turn out to be wrong and there are really no more ivory bills (which is still possible), the unlucky biologist could waste his or her life in a useless enterprise that proves nothing. The answer to the objection is that scientists are people and sometimes they do things that are unwise. While speculations may motivate scientists, they do not really belong in science.

006.3 Science and the search for error

Popper's philosophy of science may not describe everything that scientists do. However, it is an accurate description of what they are trying to do and that goes a long way toward accounting for their behavior. Once you understand that their main focus is on trying to find mistakes, a lot of things begin to make sense.

Astronomers stayed with the Ptolemaic Model of planetary motion long after the model became cumbersome and suspect. Science makes progress by proving things wrong. The way to do that is to stay with a single established framework of assumptions and push it until it breaks. Until the Ptolemaic Model came into definite contradiction with observations, it was the only game anyone would consider playing. Aristarchus and Copernicus were mostly ignored.

The newspaper reports that the latest space experiments confirm Einstein's model of gravity. You would think that scientists would be happy to be proven right. Instead, they view the result as boring. An exciting result would be one that conflicts with Einstein's model since that would lead to new science.

An amateur scientist complains that nobody will even listen to his brand new alternative model of gravity. He figures that they all have closed minds and wish to perpetuate their own ideas. In fact, the scientists are entirely focussed on testing the predictions of the established model (Einstein's) to see if it conflicts with observation. The predictions of this particular new theory are irrelevant to that task, so it is ignored.

A professional scientist comes up with a brand new alternative model of gravity and presents it as a "test theory" that suggests new observations to test the currently accepted theory. Not only does everyone listen, but he gets a large grant to develop the theory further.

In the Creationism/Intelligent Design versus Evolution arguments most non-scientists figure that is fair for all sides of an issue to be presented. Most biologists, however are hostile to such an idea. They are entirely focused on testing the predictions of the established model, evolution, to see if it conflicts with observation. An alternative model that does not make any predictions is irrelevant to that task.

The key point here is that a scientific statement is always considered to be a target for destructive testing. Everyone shoots at the *same* target, the currently established statement, until it finally crumbles into disagreement with observation. This kind of process is very different from a debate because there is usually only one side to every question, namely the currently accepted side, the model that has so far stood up to repeated testing.

006.4 The scientific fact problem

Although scientists seldom have debates, philosophers often do. One strong objection to Karl Popper's falsificationist philosophy is that it seems to imply that there is no such thing as a scientific fact. Instead, we just have "currently accepted scientific ideas" and those are required to be tentative so that they could change tomorrow. Surely there are some scientific statements that are really not tentative and could not possibly change tomorrow. For example, the "fact" that the planet Mars has existed at least up until now. The planet might be destroyed by some astronomical disaster, but we surely do not expect to one day hear that the existence of the planet was all a big mistake. However, it must be pointed out that it has already been proposed that the existence of the planet Pluto was a mistake. (The mistake was calling it a planet.)

One way around the difficulty is to say that science produces "revisable facts." Those revisable facts are just the statements that have stood up to repeated testing and are currently accepted. When an astronomer says "The expansion of the universe starting from an initial singularity 13.7 billion years ago is a scientific fact." or a biologist says "Evolution is a scientific fact." they are using this meaning of the term "scientific fact."

The real difficulty with accepting Popper's philosophy is that it seems to leave something out. It correctly describes science as basically a search for error. However most people, scientists and non-scientists alike, assume that science is supposed to produce a kind of truth. How does a search for error ever produce truth? The scientific method is very much like a well-known description of how a sculptor produces a statue from a block of marble: He or she chips away everything that is not the statue. The question then is why does one end up with a statue and not just a pile of marble chips? What pure falsificationism leaves out is the assumption that we live in a universe with fixed rules that we can discover. We can never be sure that we have the right answer to a scientific question, but we always have faith that there is a right answer.

006 Spot Check

Here are some questions to check your understanding of the material in module 006. Both the answers and where to find these questions at the website may found at the end of the Study Guide.

1 Einstein's Theory of Relativity has passed every observational test for over 100 years. Among other things, it predicts that no material object can go faster than the speed of light. Fred Zveistein (twice as smart as Einstein) has a new theory that predicts that some material objects can go faster than light. Fred presents his new theory at a scientific meeting. Which of the following would be the most likely reaction of the scientists at the meeting to this new development?

a. There is great disinterest because there is no need to replace a theory that has passed every observational test. Nobody at all comes to Fred's talk.

b. There is great hostility because the new theory challenges the established theory, so lots of scientists come to Fred's talk to debate the issue.

c. There is great interest because the new theory proves that Relativity is wrong, so lots of scientists come to Fred's talk to congratulate him.

2 Which of the following statements is a speculation?

a. All of the fish in Lake Nyak are green.

b. There are fish in Lake Nyak.

c. All of the fish in Lake Nyak are beautiful.

3 Which of the following statements is falsifiable?

a. There is intelligent life on other stars.

b. There is cheese on the Moon.

c. The Moon is made entirely of cheese.

d. Isaac Newton was the greatest scientist.

e. There is beauty in a sunset.

4 The idea of a scientific fact presents difficulties for the falsificationist philosophy that we have been discussing because that philosophy insists that every scientific statement is

a. subject to change.

b. infallible.

c. just a guess.

007: Model Building

007.1 Scientific models a.k.a. scientific theories

So far I have avoided using the term "Scientific Theory" because it is enormously misleading. A "theory" is usually thought of as a guess that is not connected to reality at all. What scientists actually do is produce *models* that *represent* real systems. The models consists of things, either real or abstract, that can be manipulated and analyzed to reveal relationships that apply to the real system. Whatever kind of model is used, the crucial feature is that it make predictions that correspond to reproducible observations. Whenever this correspondence fails, the model is either revised or discarded.

Examples

The model might be an actual miniature physical system. Suppose that you want to re-arrange the furniture in a room. Instead of moving the furniture many times, make small models of the room and each piece of furniture using rectangles of paper with the right dimensions. Then you can move these paper furniture models around in the room model any way you want and see which way things will work best. You end up with a prediction of what arrangement will work. If it turns out that the actual furniture does not fit in the predicted way, you need to correct the sizes of some of your paper furniture models or add new features to them (like door swing areas) before trying again.

The ancient Greeks also produced physical models to represent the Earth and the heavens.

This device is called an *armillary sphere.* It has moving parts that correspond to the celestial equator, the Sun, and the Moon.

The model might consist of abstract mathematical objects. The Greeks were fascinated by geometry, so they tended to construct geometrical models. As we have seen, the Earth is well modeled by a sphere.

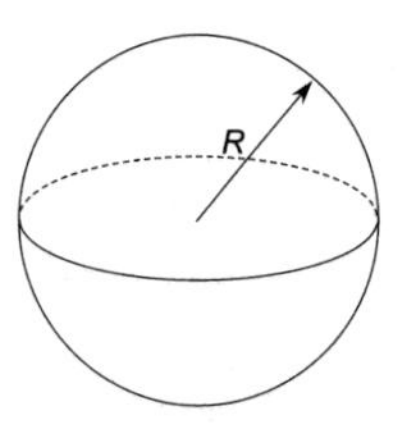

As mathematicians devise more elaborate systems of abstract objects, scientists use those objects to construct their models of reality. Oddly enough, the most modern branches of theoretical physics are once again using geometrical models to represent the real world. However, it is a geometry (differential geometry) that has evolved far beyond anything that the Greeks would recognize.

The model might also consist of concrete mathematical objects — numbers and programs in a computer. For example, weather predictions are mostly made using computer models. Like physical models and abstract models, these computer models are also subject to constant testing and are revised or discarded when they fail to predict what is observed.

007.2 Modeling the Earth-Moon System

We have seen that the ancient Greeks got into trouble when they tried to model the planets. However, they did an excellent job with the Sun, Earth, and Moon. We got this picture as part of the 30 billion dollar Apollo program.

The ancient Greeks had already pictured it in their minds 2500 years ago: A spherical Moon orbiting a spherical Earth, both lit from one side by the Sun.

Solar Eclipses

The geometrical model of what is happening in a total solar eclipse looks like this:

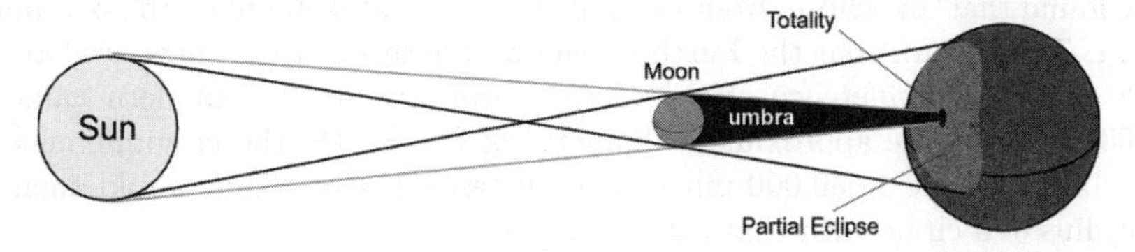

The region in which a total eclipse is seen is in the Umbra of the Moon's shadow, where no light from the Sun can reach. The observed fact that the Sun and Moon have the same angular size in the sky means that there is only a very small region on the Earth' surface where the entire Sun is covered up by the Moon. From the geometrical model, the Greeks could see that the Sun had to be larger than the Moon with a size that is exactly proportional to its distance from the Earth.

Lunar Eclipses

Here is a sequence of photographs of a lunar eclipse:

Notice that the shape of the Earth's shadow, particularly just before and after totality, is curved with a radius a few times the radius of the Moon. That was important supporting evidence for a spherical Earth that is larger than the Moon. The geometrical model of what is happening looks like this:

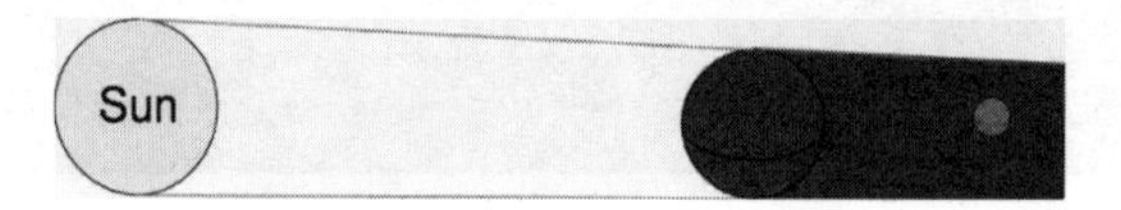

By timing the initial phase of the eclipse and the totally eclipsed phase, the Greeks could figure out that the Moon is passing through a shadow about four times its size. From the geometrical model, that meant that the Earth must be about four times the size of the Moon. Since they knew the size of the Earth, they could then figure out the size of the Moon.

The timing of the eclipse gave them another important piece of information. They found that an eclipse, from beginning to end, takes about 1/100 of a lunar month. That meant that the Earth's shadow, whose size they knew, was about 1/100 of the circumference of the Moon's orbit. Using the modern value of 15,800 miles for the approximate diameter of the Earth, the circumference of the orbit would be 1,580,000 miles. The distance to the Moon would then be the radius of a circle with that circumference:

$$r = \frac{1580000}{2\pi}\text{miles} = 251,500 \text{ miles}$$

The distance actually varies between between 220,000 miles and 252,000 miles, so this result is pretty close to being correct.

007.3 Modeling the Sun

The most important thing that the Greeks needed to know about the Sun was how far away it is. From that they could calculate how big it is and complete their geometrical model of the Sun-Earth-Moon system. The situation they decided to focus on is the phase of the Moon that we call a "quarter Moon." In that phase, we see half of the Moon in sunlight and half in darkness with the boundary between them exactly straight. The geometrical model of these two situations looks like this:

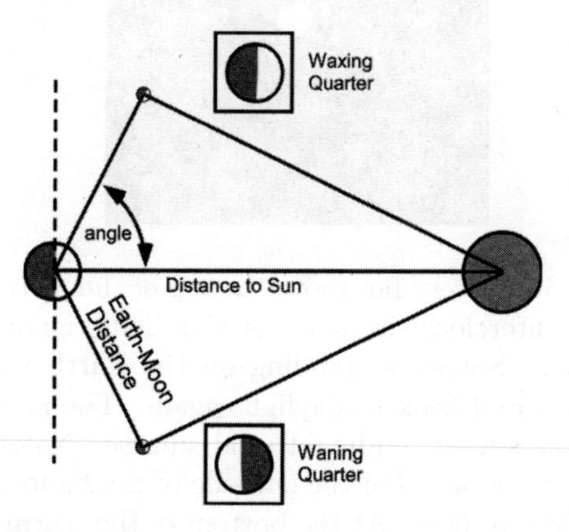

The method used by Aristarchus took advantage of the fact that both the Moon and the Sun are visible in the sky at the same time during these quarter phases, so it was possible to measure the angle between them. That is still not easy because you have to estimate the exact instant when the Moon is at the quarter phase (called the "dichotomy"). Aristarchus found that the angle is definitely larger than 87 degrees. The result was that the Sun must be at least 19 times as far away as the Moon and at least 19 times as large as the Moon.

The actual value of the angle that Aristarchus tried to measure is $89\frac{5}{6}$ degrees and the Sun is actually about 400 times as far away from the Earth as the Moon and 400 times as large. However Aristarchus did get the basic idea that the Sun is a lot farther away than the Moon and is definitely much larger than the Earth. It was that realization that led him to conclude that the established Greek model of the larger Sun orbiting the smaller Earth was probably wrong.

007.4 Modelling Local Time and Compass Directions

Start with a picture of the Earth, as seen from above the North Pole.

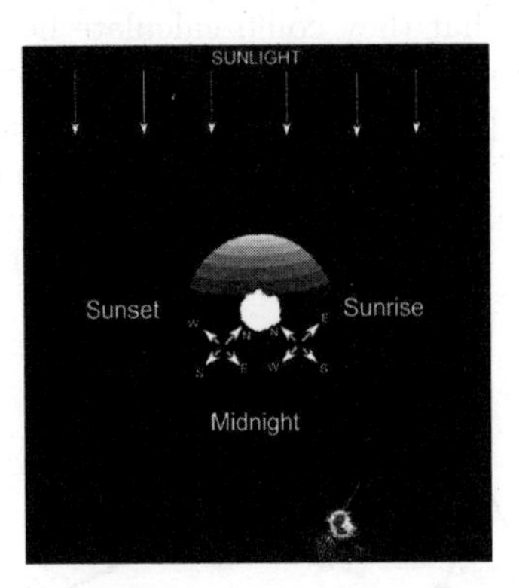

Assume that the Sun is very far above the top of the picture. In this picture, the Earth rotates counterclockwise and, for simplicity, ignore the fact that the rotation axis is tilted. Someone standing on the Earth will be carried from daylight into darkness and back to daylight again. The figure indicates where they would experience sunset, midnight, and sunrise. Notice that the time of day depends on where you are. On the left side of the figure, where it is sunset, the time might be close to 6pm. At the bottom of the picture it would be close to midnight. At the right side of the figure, where it is sunrise, the time might be close to 6am. These times will not be exact, partly because we have ignored the axis tilt that causes days to be longer in the summer and shorter in the winter and partly because these are solar times and the time that is read on clocks (legal time) agrees with solar time only at certain places in each legal time zone.

The compass directions, North, South, East, and West, also depend on where you are standing on the Earth. North is always toward the North Pole, indicated in the picture by an ice cap. West is clockwise and East is counterclockwise. The picture shows the compass directions for two locations on the Earth. Check that, when these compasses are carried around by the Earth's rotation, the Sun will set in the West and rise in the East.

007.5 Modelling the Phases of the Moon

Now enlarge the picture by including the Moon in its orbit around the Earth. It would be very difficult to do that to scale since the distance to the Moon is about 30 times the radius of the Earth. Thus, we put the Moon a lot closer in than it should be. That discrepancy in the picture needs to borne in mind when we figure how the Moon should look from the Earth.

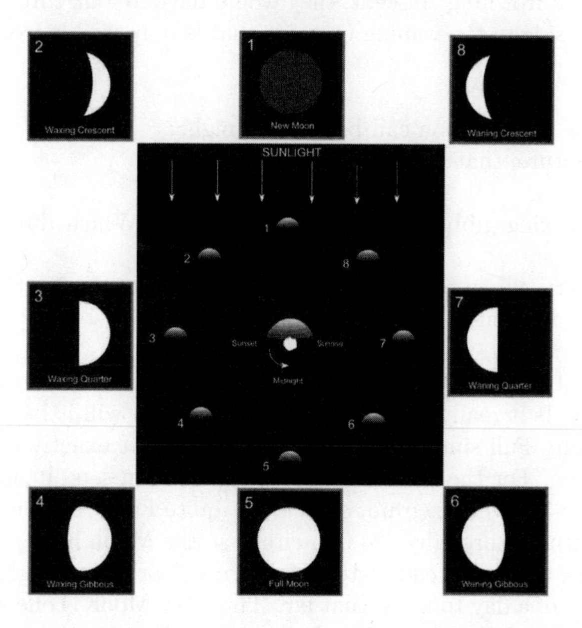

The numbered boxes indicate how the Moon looks in each stage of its orbit. The only place on Earth where the phases would be the way they are drawn here would be at the North Pole. From other places, the phases would be tilted because we are standing on a sphere. For example, from the Earth's equator one would see a waxing crescent moon that looks like a smile.

Relative to the distant stars, the Moon moves once around the Earth in about 27 days. That is called a *sidereal month.* Because the Earth-Moon system is orbiting the Sun at about one degree per day, the direction of sunlight shifts by about 27 degrees during a sidereal month, so the Moon must move through another 27 degrees to get back to the same place in relation to the direction of sunlight. As a result, a month of lunar phases (called a *synodic month*) is a bit longer than a sidereal month, about 29.5 days.

This model of the Moon's phases can be used to answer fairly complicated questions about the time of day and the direction in which each phase can be seen.

Examples

You look up and see a crescent moon that looks like an upside-down bowl or maybe a frowning mouth. Where are you? What phase is it? What time is it?

What direction are you looking?

Wherever you are, it is not this planet. Imagine someone standing on the equator near sunset in the picture. At the waxing crescent phase, they see the lit portion of the Moon in the direction of their feet, so they see a smiling crescent. Now put them on the equator near sunrise in the picture. During the waning crescent phase, they still see the lit portion near their feet and still see a smiling crescent. To see a frowning crescent, they would have to look through the Earth from the sunset side to the waning crescent that is only visible from the sunrise side.

—

Which phase of the Moon can be seen all night?

From the picture, that would be the Full Moon.

—

You see a waxing gibbous Moon at about 9pm. Which direction are you looking?

From the picture, you are looking directly away from the North Pole, so you are looking directly South.

—

You see what appears to be a full Moon rising in the East an hour before the Sun has set. Is it really a Full Moon? If not, when will it be Full?

It is not really Full since a Full Moon would rise at exactly the same time that the Sun sets. For the Moon to rise before the Sun sets, it must be before position number 5 in the diagram, so it is not quite Full. An hour corresponds to the Earth turning through 1/24 of a circle, so the Moon has 1/24 of its orbit still to go. Since the Moon completes a full synodic orbit in roughly 30 days, it takes $30/24 = \frac{5}{4}$ of a day to move that far. Thus, the Moon is one and a quarter days short of being Full. It would be Full on the following night.

—

A witness at a murder trial says that he saw the crime easily because it was a moonlit night. The crime allegedly took place at 9pm and the Moon was in its waning quarter phase. Does his story make sense?

No it does not since the waning quarter phase could not possibly be seen before midnight. There would have been no Moon in the sky at all at 9pm.

007.6 Modelling the Phases of Venus

The Ptolemaic model of the solar system had two of the planets, Mercury and Venus, moving along epicycles centered on a line that joins the Sun and the Earth. That arrangement was needed to reproduce the observation that these two planets always stay near the Sun, sometimes rising just before the Sun as "morning stars" and sometimes setting just after the Sun as "evening stars." Since these planets are spherical objects, they should show phases, just like the Moon does. The Ptolemaic model, which puts these planets between us and the Sun, would predict that there should never be a "full Mercury" or a "full Venus." The Copernican model, on the other hand, has these planets moving around the Sun and would predict that they should show all of the phases.

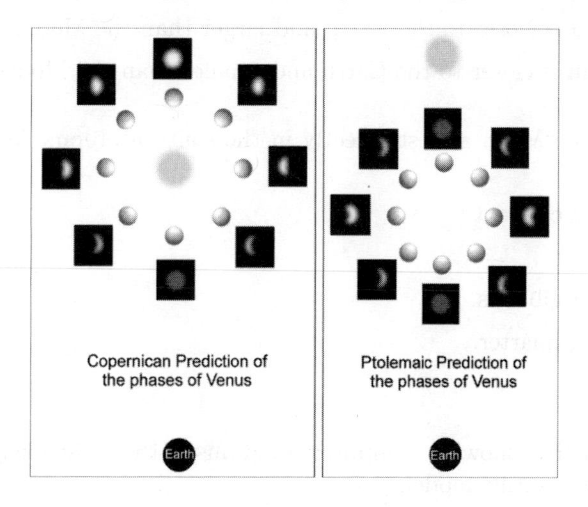

These predictions were put to the test by Galileo. He had heard that two lenses could be put together to make a device for seeing things at a distance, a *telescope*. It is still not totally clear who first discovered the telescope. Hans Lipperhey applied for a patent in the Netherlands in 1608, but was denied because it is just too easy to put two lenses together. By the end of 1609, telescopes of about 3x magnification were being sold all over Europe. Galileo constructed a whole series of telescopes, improving the design each time. Eventually he produced a 20x telescope and turned it on the heavens. He saw that Venus indeed had phases and they were the ones predicted by the Copernican model.

It took several years for the rest of Europe to catch up to Galileo in telescope design so that they could reproduce his observations. Once they did, the Ptolemaic model of the heavens had its first major conflict with observation and, 66 years after the death of Copernicus, the Copernican System had its first big break. Of course, 1609 was also the year that Kepler's *Astronomia Nova* was published, and the original Copernican System was replaced by Kepler's Laws of Planetary Motion.

007 Spot Check

Here are some questions to check your understanding of the material in module 007. Both the answers and where to find these questions at the website may found at the end of the Study Guide.

1 From the way that solar eclipses happen, the ancient Greeks concluded that

a. the Sun is farther from the Earth and smaller than the Moon

b. the Sun is farther from the Earth and larger than the Moon.

c. the Sun is closer to the Earth and larger than the Moon.

d. the Sun is closer to the Earth and smaller than the Moon.

2 You see the Moon almost directly in the south at 10pm. What phase is it?

a. Waxing crescent.

b. New.

c. Waxing gibbous.

d. Waxing quarter.

e. Full.

3 Which of the following computer programs is the most likely to be considered a scientific model:

a. A simulation game in which people can spend a day at the beach.

b. A simulation,using currently accepted physical laws, of waves crashing on the beach.

c. An animated screen-saver that shows waves crashing on the beach.

4 If you are looking down over the north pole of the Earth, you will see the Earth rotate

a. clockwise.

b. upward.

c. counterclockwise.

d. downward.

5 The first telescope was (possibly) built by

   a. Galileo.

   b. Newton.

   c. Tycho Brahe.

   d. Someone in the Netherlands.

6 The ancient Greeks were able to estimate the distance from the Earth to the Sun by using

   a. lunar eclipses.

   b. the quarter phases of the Moon.

   c. the observed sizes of the Sun and Moon in the sky.

   d. solar eclipses.

008: Models of Motion

008.1 Aristotle

The ancient Greek ideas about motion are usually attributed to Aristotle. They characterized motion only by its *speed*. Earthly or mundane objects were naturally at rest. Any time that such an object was moving, there needed to be something pushing it. The speed would be proportional to how hard it was pushed. Heavenly objects were different. They kept moving because there was no reason for them to stop.

Projectiles, such as arrows or spears, keep moving long after they are launched. Aristotle accounted for this behavior by saying that the air displaced from the front of the object moved around to the back and pushed it.

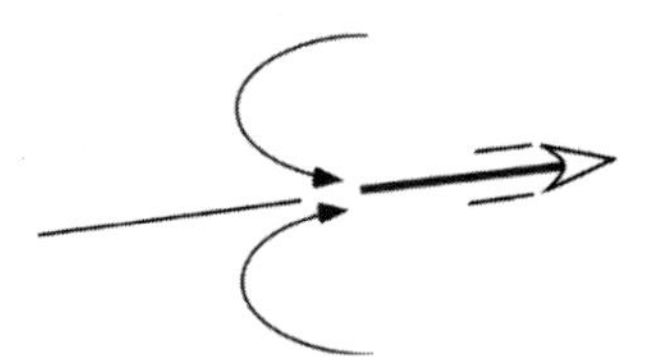

He predicted that a spear thrown in a vacuum, where there is no air, would simply stop the instant it left the thrower's hand.

Gravity was assumed to be a force pulling every mundane object toward the center of the Earth. Since speed was assumed to be proportional to force, Aristotle predicted that heavier objects would fall faster than lighter objects. For a wooden ball and an iron ball, both the same size to make the effects of the air equal, and both dropped at the same time from the same height, Aristotle would predict a motion like this:

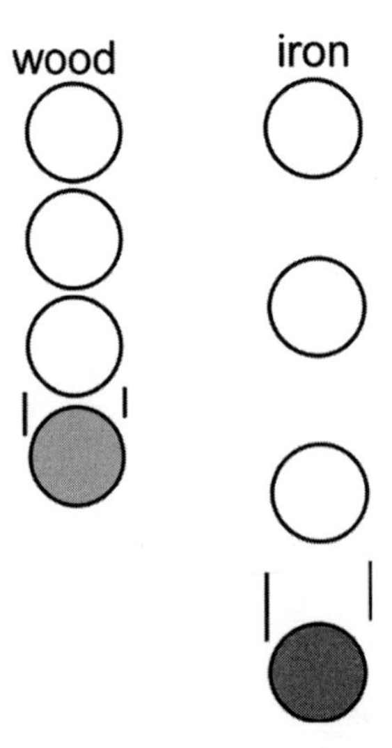

The picture shows a series of snapshots at equal intervals. Of course the ancient Greeks did not have flash cameras, so they could not check this prediction in detail. However they could easily have noticed which object hit the ground first. However, the idea of setting up such an "artificial" situation to test a set of logical conclusions about the real world would not have seemed reasonable to the ancient Greeks.

008.2 The Law of Inertia

Galileo liked to build things (telescopes for example) and realized that he could gain an understanding of real but complicated and difficult to measure events by building things that were simple and easy to measure. The ancient Greeks, with their rigid separation of practical technology from philosophy, would never have understood. To understand motion, Galileo found a way to make it simple, slow, and easy to measure: He rolled things down inclined planes.

The experiments were probably done between 1589 and 1592 while Galileo was writing a book, *De Motu*, that he never actually published. One set of experiments involved facing inclined planes, like this:

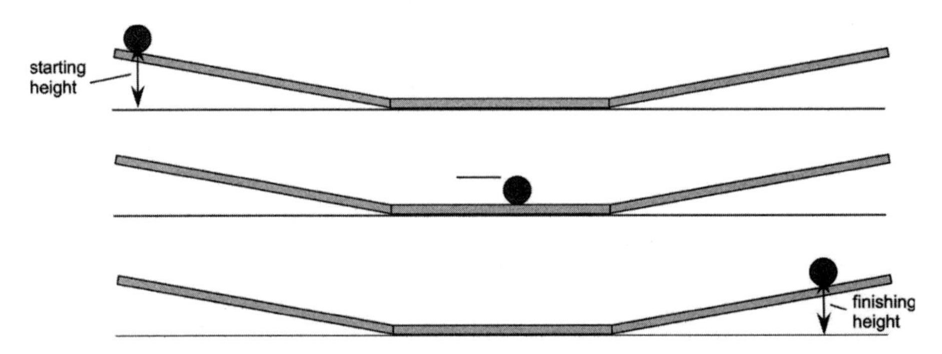

By using this arrangement, Galileo could avoid having to measure the speed of an object. He only had to measure the height of the ball at its starting and stopping points. He noticed that making the flat part of the arrangement longer caused the finishing height to decrease, so the ball really was slowing down. However, that slowing was not acting as Aristotle's picture would imply. The displaced air did not seem to be helping, for one thing. It is not clear if Galileo did the experiment, but it is easy to compare a large cork ball to a small metal one, both weighing the same. The large ball has more area for the air to act on, so Aristotle would predict that it should lose less speed. In fact, it loses more. One experiment that Galileo did report on is to make both the planes and the balls smoother. In that case, they lose less speed. Aristotle would predict that there would be an inherent and unavoidable loss of speed. Galileo found no such thing and concluded that the slowing was entirely due to friction forces that could, in principle, be reduced. He concluded that

> *In the absence of outside forces, an object at rest will remain at rest and an object in motion will continue at constant speed in a straight line.*

008.3 Acceleration

Galileo did not have anything like a stopwatch, so it was difficult for him to measure the speeds of real objects over very short time intervals. In characteristic fashion, he made it easier by considering long inclined planes with shallow slopes. He had figured out that a pendulum takes the same amount of time for each swing, so he did have a way of marking equal time intervals. The results looked like this:

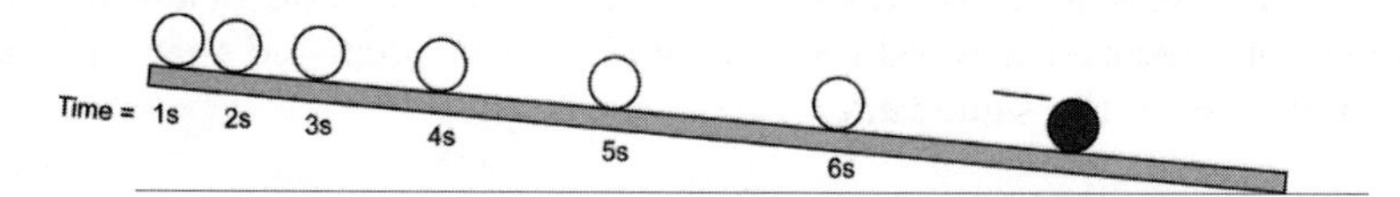

This experiment can be done in several ways. One way is to have someone call out the times and just make marks on the side of the plane where the ball is. A more accurate way is to grab and stop the ball when a time is called out so that its position can be measured accurately. The result is that the ball goes farther during each time interval — Its speed increases at a constant rate. The rate of increase of the speed is called the *acceleration.* The concept of acceleration was one that the ancient Greeks had never considered.

Galileo mostly considered objects that were moving in a straight line. For the general case, where objects might be moving in changing directions, we need to define acceleration as the rate of change of velocity, where *velocity* has both size (the speed) and direction. In terms of this concept, the Law of inertia simplifies to just this:

> *In the absence of outside forces on an object, the acceleration of the object is zero.*

008.4 The Universality of Free-fall

Galileo noticed that it did not seem to matter how heavy the balls were or what they were made out of. For a given incline, their acceleration was nearly the same. That led him to predict that freely falling objects would behave the same way. He could not measure the acceleration of falling objects because he did not have precision instruments. However, he could do the experiment that the Greeks did not do. He dropped two balls, both the same size, one of wood and one of iron. As we saw, Aristotle would predict that the iron ball would hit the ground long before the wood one. Galileo's theory predicted that the balls would hit at exactly the same time.

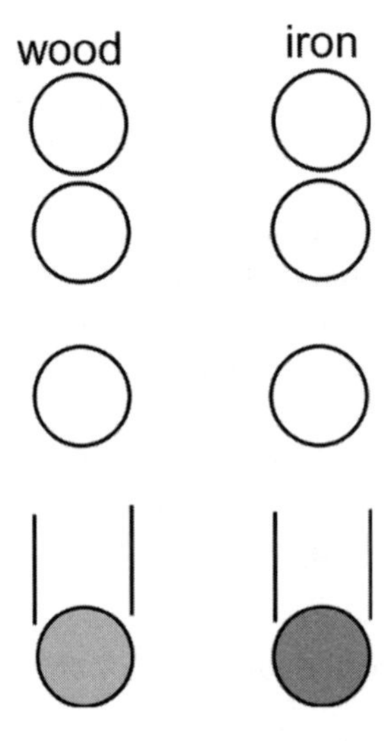

Galileo did not have timed photoflashes to take a series of pictures like the ones shown here, but he could see if the balls hit the ground together.

Actually they did not hit at exactly the same time. The balls did hit closely enough to rule out Aristotle's prediction. Depending on how far the balls had fallen, sometimes the wood one would hit first and sometimes the iron one. Like an honest experimenter, Galileo reported exactly what really happened. Modern re-enactments of the experiment get the same discrepancies, so we can be sure that Galileo really did do the experiment. Evidently human reaction times favor letting go of the wood ball before the heavier iron one while air friction favors the iron ball in a long drop. Galileo's conclusion is called the *Universality of Free Fall:*

> *Objects in free-fall (i.e. acted upon only by gravity) accelerate downward at a rate that is the same for all objects.*

At the surface of the Earth, the rate is 9.8 meters per second per second.

008.5 Force and Mass

From Galileo, Newton took the idea that it is acceleration and not speed that responds to outside forces. He made Galileo's Law of Inertia, his

First Law of Motion: *When the total force on an object is zero, its acceleration is also zero.*

Newton then considered the case where the acceleration of an object is not zero. That could only happen if there is some force acting on the object. The simplest assumption about such a force is that it acts in the direction of the acceleration that it is producing and is proportional to the amount of acceleration: Twice as much force, twice the acceleration and so on. That became his

Second Law of Motion: *If a total force F on an object produces an amount of acceleration a, then these quantities are related by the equation*

$$F = ma$$

where m depends only on the object and not on F or a.

The quantity m in the second law is called the *mass* of the object. The second law equation makes the mass a measure of how difficult it is to make an object accelerate. An object with larger mass needs more force to produce a given acceleration. What is not often pointed out about the second law is that it is useful mainly because of two important observations about Nature:

The Addition Laws: *If two independent forces F_1 and F_2 act on an object, the total force on the object is the sum $F_1 + F_2$. If two objects with masses m_1 and m_2 are connected so that they move as a unit, they form an object of mass $m_1 + m_2$.*

These addition laws make it possible to measure forces and masses with spring and weight balances and make mass a good measure of how much material there is in an object. Without them, Newton's Second Law is an empty definition.

The unit of mass is still a physical object stored in Paris. It is called a kilogram.

The unit of force is based on the kilogram through Newton's second law: The force that causes a one kilogram mass to accelerate at one meter per second per second is called a Newton. In familiar units, a Newton is about a quarter of a pound: The weight of a hamburger.

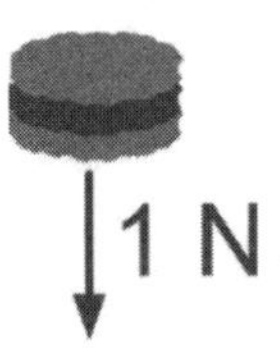

Example: Addition of Forces

Two people push on a car. One person pushes it forward with a force of 150N while the other pushes it forward with a force of 100N. What is the total force they exert on the car? Suppose that they push in opposite directions, with one pushing the car forward with a force of 150N while the other gets in front of the car and tries to push it backwards with a force of 100N. Then what is the total force they exert on the car?

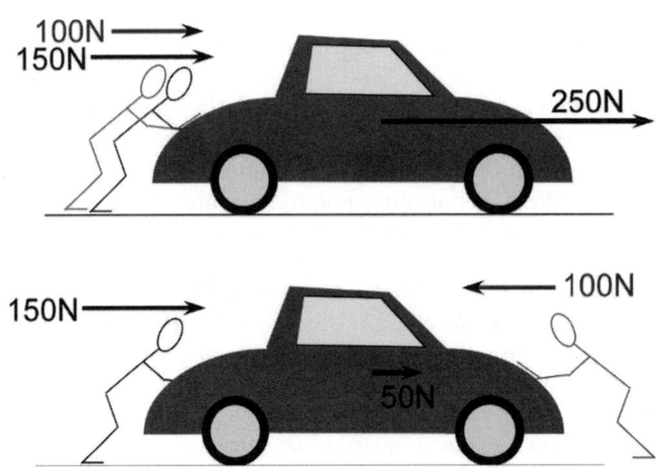

Notice that you have to take direction into account when adding forces. In the first case, the forces add while, in the second case, they subtract.

Example: Ion Drive Spacecraft

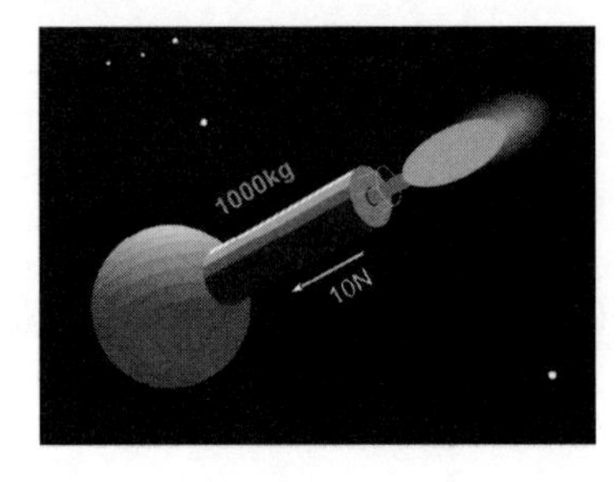

An ion drive uses electric fields to accelerate charged atoms to very high speeds to generate rocket thrust. The thrust is usually very small but can be maintained for a very long time. If the probe shown here fires its ion drive for one year, how fast will it be going?

First find the acceleration of the probe. The force on the probe is 10 N and its mass is 1000kg, so

$$a = \frac{F}{m} = \frac{10\,\mathrm{N}}{1000\,\mathrm{kg}} = 0.01\,\mathrm{m/\,s^2}$$

which means that the probe will gain one hundreth of a meter per second of speed every second. The number of seconds in a year is about $3.14 \times 10^7\,\mathrm{s}$ so we multiply to find the speed after a year: $0.01\,\mathrm{m/\,s^2} \times 3.14 \times 10^7\,\mathrm{s} = 3.14 \times 10^5\,\frac{\mathrm{m}}{\mathrm{s}}$ which works out to 314 kilometers per second.

008.6 Action and Reaction

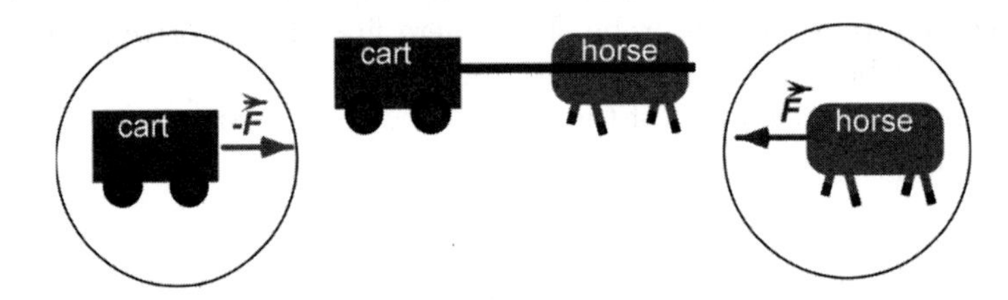

Third Law of Motion: *If object A exerts a force on object B, then object B exerts an equal and opposite force on A.*

Newton's third law of motion places an extreme limitation on the forces that can act on objects. For every force that is exerted on an object, there is *another* object that is exerting that force and that object feels an exactly equal and opposite force. This law holds even if the objects are moving and accelerating. One major consequence of it is that internal forces, forces between different parts of the same system, will always cancel out in the sum of all the forces on the system. That is why you cannot lift yourself up by your own bootstraps:

Example: The Horse and Cart System

You might wonder how the horse and cart pictured above can actually go anywhere if the forces cancel out. The answer is to use the law of action and reaction again, taking *another* object into account. The horse pushes back on the *road.* By the Third Law, the road must push forward on the horse. When you add up all the forces on the horse-cart system, it is the *force exerted by the road* that makes the system go.

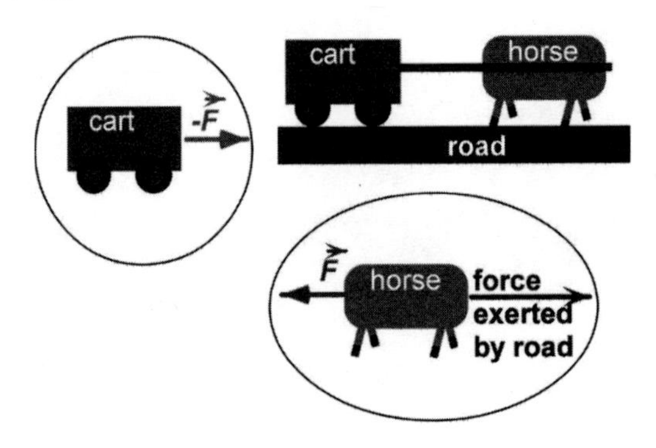

Example: How a Rocket Works

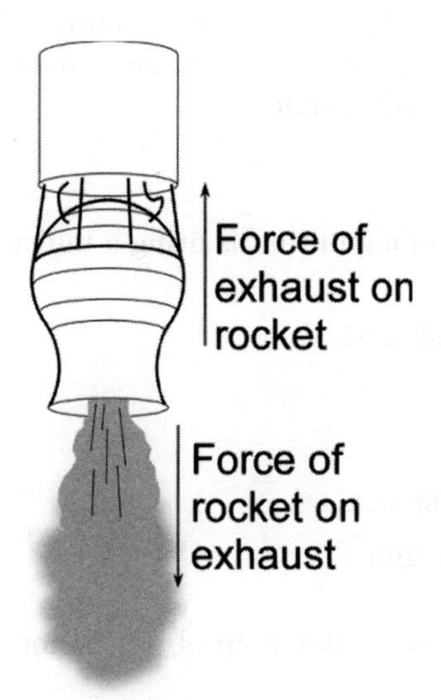

Notice that the rocket pushes against its own exhaust. It works best when there is no air around it at all. Also notice that the details of just how the exhaust pushes on the rocket do not matter. For example, the details of what happens in the ion rocket considered earlier are very different from the compressed and heated gases that push against the combustion chamber in this example. The Law of Action and Reaction does not care about such details and insists that any device that pushes any kind of exhaust out the back will feel a forward force from that exhaust.

008 Spot Check

Here are some questions to check your understanding of the material in module 008. Both the answers and where to find these questions at the website may found at the end of the Study Guide.

1 The recoil or 'kick' of a gun that is firing a bullet is a force exerted on the gun by

a. the hand of the shooter.

b. the gun itself.

c. the bullet.

d. the air around the gun.

e. the inertia of the gun.

2 The Law of Inertia says that if an object is not acted on by any outside force,

a. its acceleration will not be zero.

b. its speed will not be zero.

c. its acceleration will be zero.

d. its speed will equal its acceleration.

e. its speed will be zero.

3 Galileo's approach to understanding moving objects was to

a. observe everyday objects such as spears and horses.

b. make up imaginary situations that he could understand.

c. rely on pure logic and debate to arrive at the truth.

d. build things that he could measure.

4 What total force will cause an object with a mass of 2kg to gain 5 meters per second every second?

a. 9.8 Newtons.

b. 490 Newtons.

c. 5 Newtons.

d. 10 Newtons.

e. 2 Newtons.

5 When Galileo dropped a wooden ball and a heavier iron ball at the same time, he found that

a. the wooden ball always hit long before the iron one.

b. both balls always hit at exactly the same time.

c. sometimes the wooden ball hit first, sometimes the iron one hit first.

d. the iron ball always hit long before the wooden one.

6 Aristotle said that a moving earthly or 'mundane' object with nothing pushing or pulling on it will always

a. speed up.

b. keep moving at the same speed.

c. follow a circular path.

d. slow down and stop.

009: Models of Gravity

009.1: The Problem of Explaining Kepler's Laws

By Newton's time, the stunning success of Kepler's Laws of Planetary Motion had forced everyone to accept a Sun-centered view of the Solar System. However, one of the original objections to such a system was still unanswered: It required the static Greek theory of gravity, with everything pulled toward the center of the universe, to be abandoned. Thus, a new theory of gravity was needed. In addition, Kepler's Laws themselves were only inspired fits to observational data and needed to be explained.

One idea that began with Kepler and appeared many times during that era was that some sort of force emanated from the Sun and that force fell off with distance. Even the idea that all objects exert gravitational forces on each other was in the air of the time. However, Newton had two big advantages over his contemporaries: (1) He understood that it was the acceleration of an object that indicates the force on it. (2) he had invented what we now call calculus. a way of calculating rates of change and in particular, a way of calculating the magnitude and direction of acceleration of a planet that was following Kepler's Laws. He found that it looked like this picture:

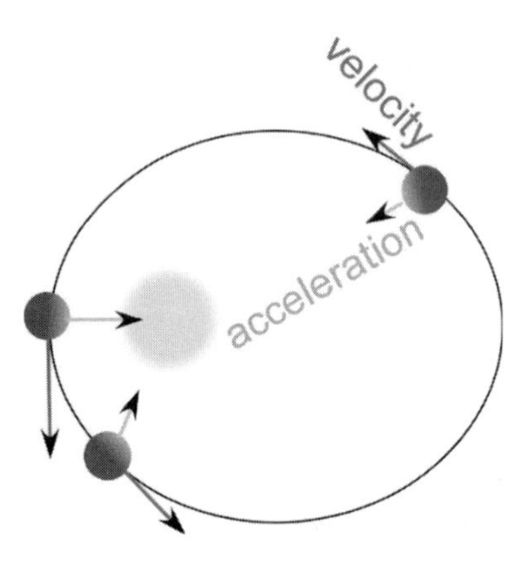

The acceleration always pointed directly toward the Sun and had a magnitude that fell off with distance in proportion to $1/(\text{distance})^2$. When the planet is at aphelion, its acceleration corresponds only to the changing direction of the planet's motion.

009.2: The Moon and the Apple

Newton was also able to calculate the direction and magnitude of the Moon in its orbit around the Earth. He found that here again, the acceleration was exactly toward the primary body, the Earth in this case.

The magnitude of the acceleration was approximately 1/3600 the acceleration of an object (like an apple) falling near the surface of the Earth. Since it was known that the Moon is approximately 60 Earth-radii away, that meant that the Moon is 60 times as far from the center of the Earth as an apple would be. Just as for the planets following Kepler's Laws, Newton found that the Moon and the Apple were accelerating toward their primary body at a rate that fell off with distance in proportion to $1/(\text{distance})^2$.

A side note: When Newton used the measured value for the distance to the Moon, he did not quite get the acceleration falling off in proportion to $1/(\text{distance})^2$. The discrepancy was not large, however, so he assumed that the measured value must be a bit off. That is an early example of a statement attributed to Sir Arthur Eddington:

> "No experiment should be believed until it has been confirmed by theory."

009.3: Newton's Law of Universal Gravitation

Newton had figured out that there was not just one center of gravitational attraction. *Everything was attracting everything else.* Consider an object with mass m and another object with mass M, separated by a distance D.

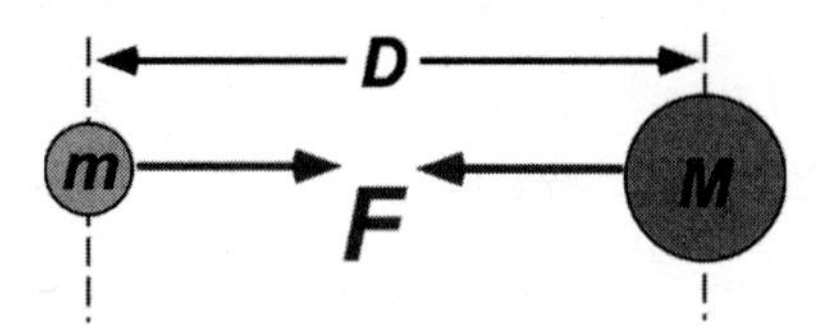

Here, M might be the mass of the Earth and m the mass of a falling apple. Since the force of gravity on an object here on Earth is proportional to the mass of the object, the force on an object of mass m had to be proportional to m. But the Law of Action and reaction would insist that the exact same force must be acting on the object of mass M, so the force must be proportional to both m and M. Combine that requirement with Newton's observation that the force falls off as the square of the distance and the only law that works is

$$F = G\frac{mM}{D^2}$$

where G is a constant number. The value of G is extremely small, which explains why one of the masses needs to be extremely large for the force of gravity to be noticed.

009.4: Unifying Physical Law

Newton's theory of gravity unified things that were previously thought to be completely unrelated:

- falling objects on Earth,
- the motion of heavenly bodies such as the Moon and planets,
- the movement of ocean tides.

Galileo's observation that all objects fall alike is explained by combining the Law of Force and Mass, $F = ma$, with the law of gravity.

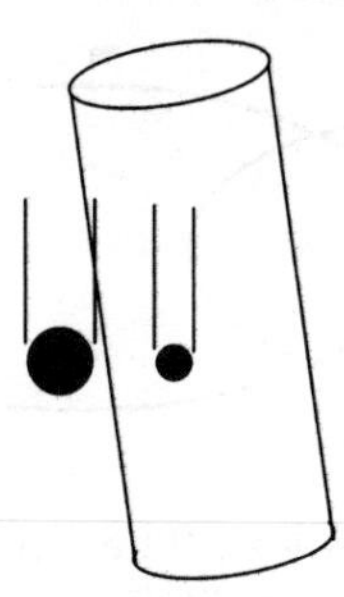

The Law of Force and Mass says that more massive objects *need* more force to accelerate at a given rate. The law of Universal Gravitation says that more massive falling objects *get* more force, so they can accelerate at the same rate.

Our Moon is actually falling toward the Earth, attracted by gravity. Similarly, the moons of all the other planets are being attracted to those planets by gravity and all of the planets are actually falling toward the Sun (without ever getting there because they go around it).

The tides occur because the Moon's gravity attracts different parts of the Earth by different amounts.

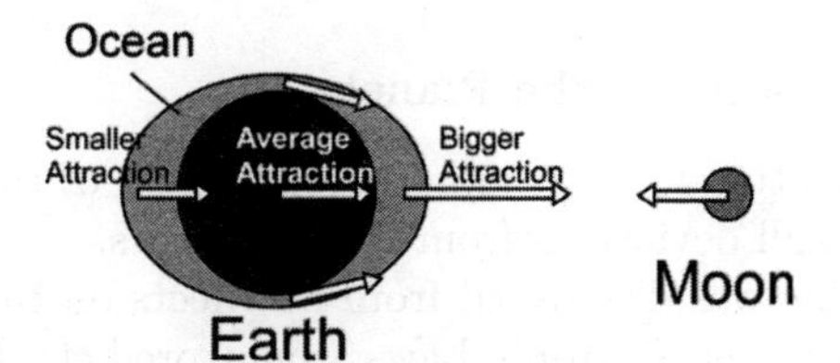

The oceans directly under the Moon are attracted more strongly than the Earth as a whole because they are closer to the Moon. The oceans directly opposite the Moon are attracted less strongly than the Earth as a whole because they are farther from the Moon.

009.5: Making New Predictions

A unified explanation of things that were thought to be different often gives rise to entirely new predictions. In the case of Newton's Theory of Gravity, there were two important corrections to Kepler's Laws of Planetary Motion:

Center of Mass Motion

Because of Newton's Law of Action and Reaction, orbiting objects accelerate towards each other. Neither object can stand still and both follow elliptical orbits with the center of mass of the system at rest between them.

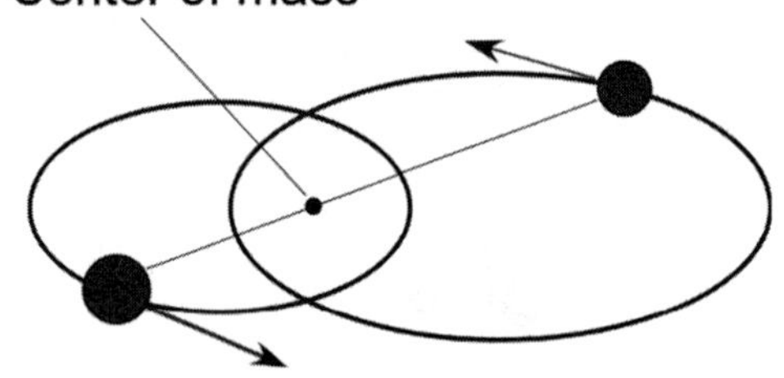

The more massive object moves less. The Sun is so much more massive than any of the planets that it moves very little in reaction to their motion. That is why Kepler's Laws work so well. It does move just a bit, however, and that slight "wobble" has been used to detect planets around other stars. Several hundred extrasolar planets have been found in this way.

When corrected for the "wobble" of the Sun, Kepler's Third Law of planetary motion is modified slightly. The quantity that is the same for every planet is actually not T^2/R^3 but

$$M_{\text{total}}T^2/R^3$$

where M_{total} is the sum of the mass of the Sun and the mass of the individual planet.

Mutual Attraction Between the Planets

Because every object attracts every other object, each of the planets attracts the others, causing small deviations from Kepler's Laws.

The planet Neptune was discovered from its effects on the orbit of Uranus. Uranus was not moving as Newton's Laws would predict. It was found that the discrepancy could be accounted for by an unknown massive planet. When telescopes were pointed at the predicted location of the unknown planet, it was indeed there.

009.6: Ballistic Missiles and Artificial Satellites

The paths of spacecraft near the Earth follow the same rules as all orbiting bodies. Their paths are ellipses with the center of the Earth as one focus.

For an object moving near the Earth's surface at less than five miles per second, the path is an ellipse that passes through the Earth.

Of course only the part of the path that is above the Earth is actually followed and the object is a ballistic missile going from a launch point to an impact point.

For a spacecraft moving horizontally at 5 miles per second near the Earth's surface (but above the atmosphere so that it does not slow down), the path is a circle.

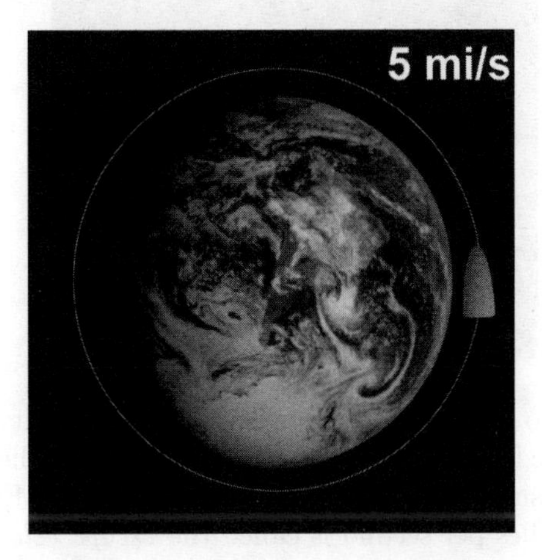

The spacecraft is accelerating toward the earth at almost the same rate as an apple falling near the ground. However it moves fast enough that the Earth curves out from under it as fast as it falls. It is an artificial Earth satellite.

If a spacecraft moves horizontally at more than 5 miles per second but less than 7 miles per second, its orbit will be an ellipse, but this time a rising ellipse that takes it farther away from the Earth.

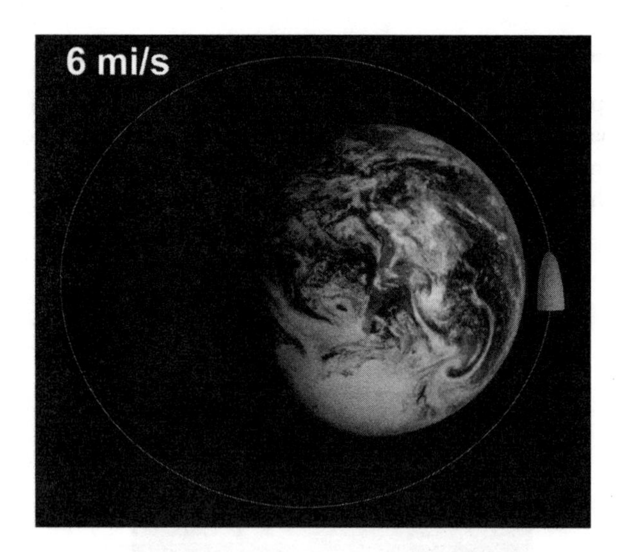

As the spacecraft moves farther from Earth, it slows down. It then speeds up again as it returns to its starting point.

Finally, if the spacecraft acquires a velocity greater than 7 miles per second near the Earth's surface,

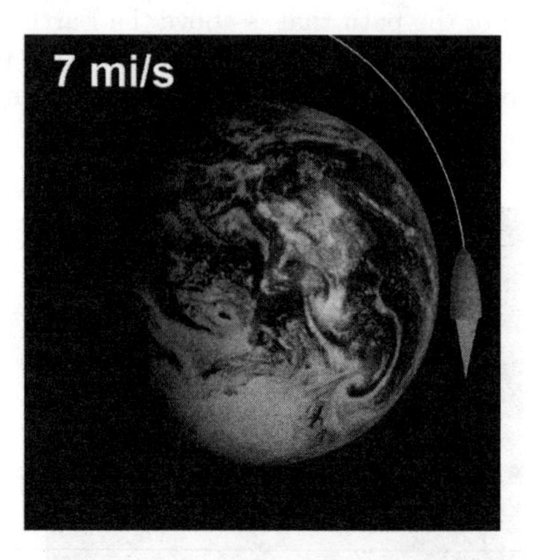

it will never return to Earth. The force of gravity slows it down as it moves away, but the distance from the Earth is increasing so that the force of gravity is decreasing and never succeeds in turning it back toward the Earth. The velocity of approximately 7 miles per second is called the *escape velocity.*

009 Spot Check

Here are some questions to check your understanding of the material in module 009. Both the answers and where to find these questions at the website may found at the end of the Study Guide.

1 In the ancient Greek theory of gravity, everything was attracted to the center of the universe. In Newton's theory of gravity, everything was attracted

a. only to massive heavenly objects such as the Sun, Moon, planets, and the Earth.

b. only to the Sun.

c. to every other object in the universe.

d. only to the center of the Earth.

2 When Newton calculated the magnitude and direction of the acceleration of Earth's Moon, he found that the direction was

a. opposite to the direction of the Moon's motion.

b. in the direction of the Moon's motion.

c. toward the Earth.

d. between the direction of the Moon's motion and the direction from the Moon to the Earth.

e. away from the Earth.

3 The International Space Station (ISS) is in a roughly circular orbit near the surface of the Earth, moving at around 5 miles per second. Suppose that it is desired to raise it to a new circular orbit, farther from the surface by having the space shuttle give it one or more short boosts. Which of the following schemes will work?

a. Push it directly upward, away from the Earth.

b. Increase its speed to 6 miles per second and then give it another speed boost when its distance from the Earth stops increasing.

c. Increase its speed to 6 miles per second to put it on a rising path.

d. Increase its speed to 8 miles per second to put it on a rising path.

e. Decrease its speed to 4.9 miles per second and then give it a speed boost when its distance from the Earth stops decreasing.

4 The force of gravity explains

a. how things fall and how the Sun shines.

b. how planets move and how the Sun shines.

c. how things fall and how lightning works.

d. how the tides and lightning work.

e. how planets move and how the tides work.

5 In comparison to Kepler's Laws of Planetary Motion, Newton's theory of Universal Gravitation predicted

a. a completely different set of motions.

b. the same motions interpreted differently.

c. almost the same motions but with corrections.

d. exactly the same motions.

010: Overview of the Solar System

010.1 The (very) Big Picture

The scale of the Solar system is so large that it is impossible to present it in just one picture. Here is our little group of inner planets:

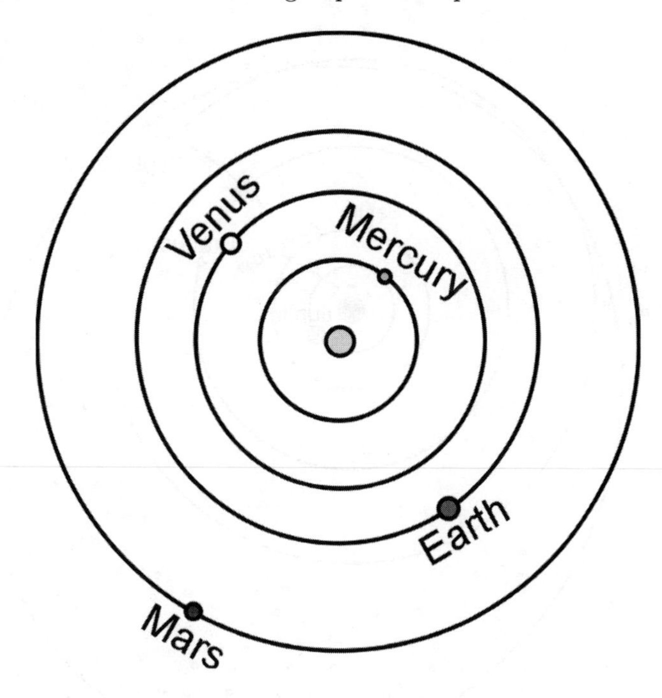

The sizes of these objects are wildly out of scale because otherwise they would be dots too small to see. The best unit of distance to use in this picture is the *astronomical unit*, the average distance from the Earth to the Sun. It is usually abbreviated au. Here are some *rough approximations* of the actual sizes of things on this scale:

- Diameter of the Sun = 1/100 au.
- Diameter of the Earth = 1/10,000 au.
- Distance from the Earth to the Moon = 1/400 au.

Think of building a true scale model of the inner solar system. Put the Earth 10 meters from the Sun, so our scale is 1au = 10m. That way we can get the Earth in the same large room as the Sun. The Sun is then 10cm in diameter, or about the size of an orange. The earth, is on the far side of the room and has a diameter of just one millimeter, about the size of a grain of rice and has an even tinier Moon orbiting about 4 centimeters away from it. The other inner planets are somewhat smaller grains of rice scattered around the room. Even this inner region of the Solar System is mostly empty space.

Moving up in scale, here are the outer planets and the region just beyond

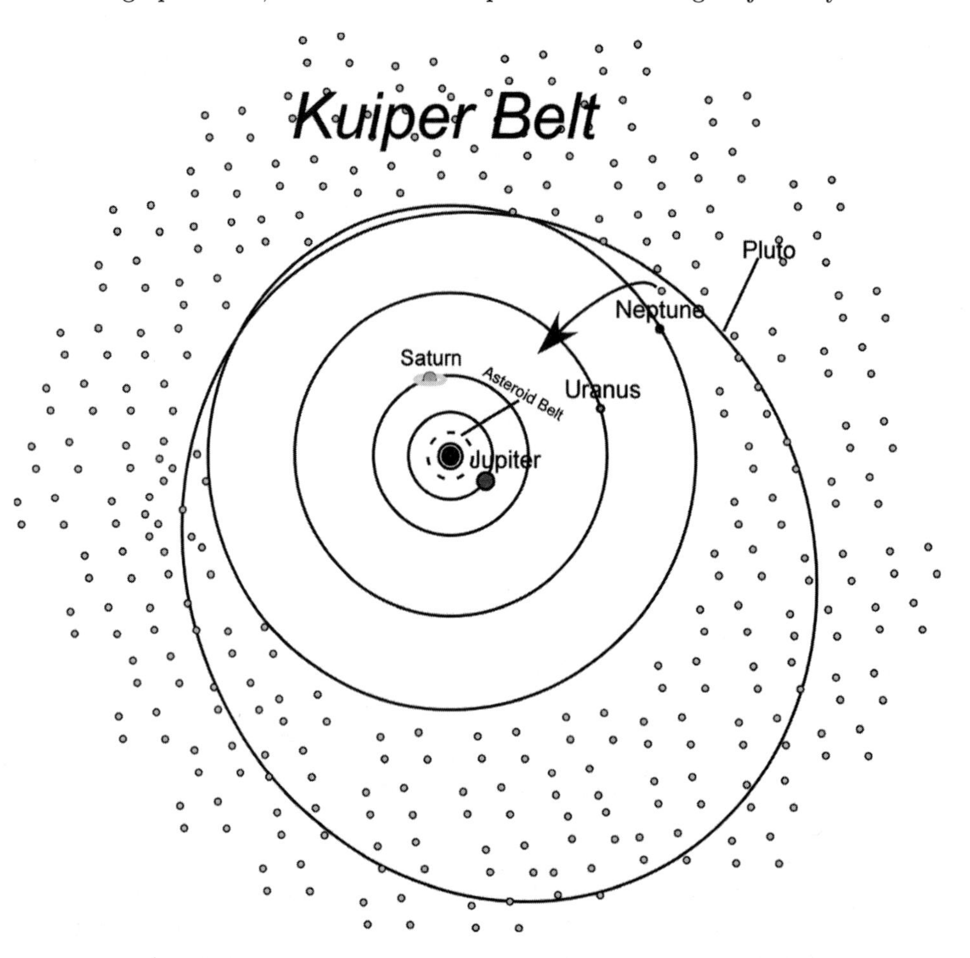

In this picture, you can just make out the orbit of Mars inside the *asteroid belt*. The farthest real planet from the Sun is Neptune, at a distance of about 30au from the Sun. Beyond Neptune is the *Kuiper Belt*, which contains dozens of *dwarf planets* such as Pluto. Beyond that, and not shown here, is the largely unexplored Oort cloud, extending to 50,000 au. Beyond that, the nearest star is 253,000 au from our Sun.

010.2 The Terrestrial Planets

The terrestrial or Earthlike planets:

are small.

are made of high density material (rock and iron).

are close to the Sun.

have solid surfaces.

have no moons (of their own).

do not have rings.

Density and Interior Structure

Density, the mass of an object divided by its volume, gives a useful clue to what an object is made of. It is an easy number to calculate for the planets since you just need the total mass, which can be gotten from Newton's Law of Universal Gravitation, and the volume, which can be calculated from the size. Here are the densities of some common materials and the densities of the terrestrial planets:

Material	Density
water	$1000 kg/m^3$
granite rock	$2700 kg/m^3$
iron	$7800 kg/m^3$

Body	Density
Mercury	$5400 kg/m^3$
Venus	$5200 kg/m^3$
Earth	$5500 kg/m^3$
Mars	$3900 kg/m^3$
Moon	$3300 kg/m^3$

From these densities, you can conclude that all of these planets are made of more than rock and water. The Earth is known to have a large iron core, so it is reasonable to infer that Mercury and Venus do also. Mars is significantly less dense than the others, so it probably has a smaller iron core than they do and the Moon would have a still smaller core. The magnetic fields of these bodies originate in their iron cores and provide still more evidence of what their interiors are like.

Here are the relative sizes and suspected interior structures of the four terrestrial planets as well as our own Moon.

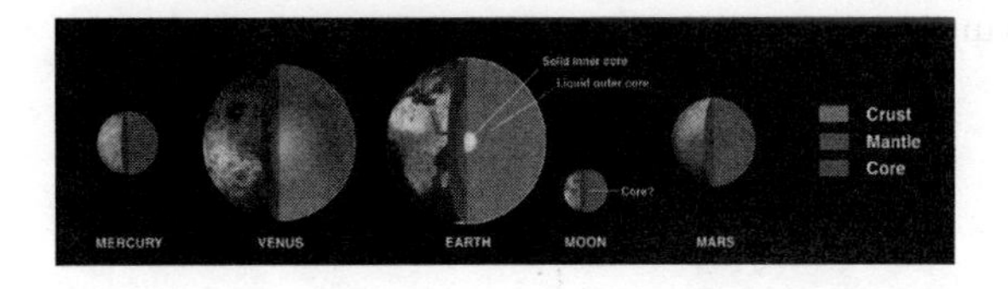

Notice that Venus and Earth are near twins in size and interior structure.

010.3 The Jovian Gas-giant Planets

The Jovian planets:

are large.

are made of low density material (gas and ice).

are far from the Sun.

lack solid surfaces.

have extensive systems of moons.

all have rings.

Density and Interior Structure

Here are the densities of some common materials and the densities of the Jovian planets:

water	$1000 kg/m^3$
granite rock	$2700 kg/m^3$
iron	$7800 kg/m^3$

Neptune	$1600 kg/m^3$
Uranus	$1300 kg/m^3$
Saturn	$700 kg/m^3$
Jupiter	$1300 kg/m^3$

These planets consist mostly of gas and ice with very small rocky cores. Saturn would float on water, if you could find a large enough body of water to put it in!

Here are the relative sizes of the four Jovian planets compared to Earth and to each other:

Although only Saturn has rings that can be seen in the picture, all of these planets have ring systems.

010.4 Asteroids

The inner solar system contains a number of rocky objects that are far too small to be planets. In fact most of them, like Ida here are too small to be spherical.

The asteroid 243 Ida is 58km long and, oddly enough, has a 'moon' orbiting it. The picture was taken by the Galileo spacecraft (August 28, 1993). Ida's moon, Dactyl is 1.2 by 1.4 by 1.6 km in size.

Most of the asteroids orbit in the Asteroid Belt, between the orbit of Mars and the orbit of Jupiter. It is thought that Jupiter's gravity kept the material in that region too stirred up to form a planet. The two largest asteroids in the Belt are Ceres and Vesta.

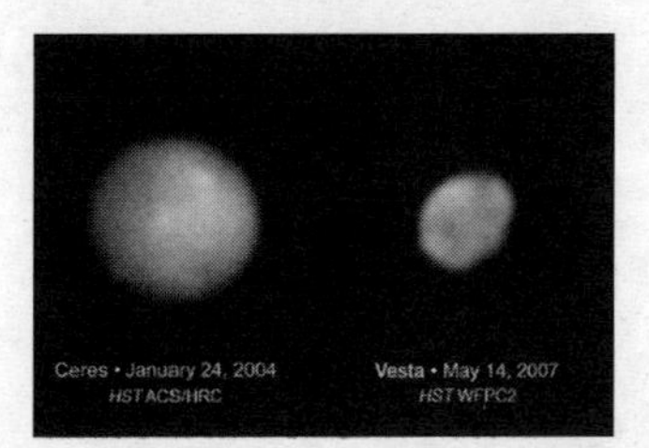

As you can see from these Hubble Space Telescope images, they are large enough for gravity to have pulled them into spherical shapes. Consequently they are now regarded as *dwarf planets* rather than asteroids.

Some asteroids have collected near stable points in the orbit of Jupiter, 60 degrees ahead and behind the planet. The Eastern Trojan Asteroids are ahead of Jupiter while the Western Trojan Asteroids lag behind it.

A few asteroids have gotten into orbits that cross the orbit of the Earth. These occasionally hit the Earth with consequences similar to a major thermonuclear war.

010.5 Comets

Have orbits that often carry them far beyond any of the planets and are made mostly of ice and frozen gas. As they approach the Sun on long elliptical orbits, the frozen gas and ice begin to vaporize, sending out jets of material as in this photo of Halley's Comet taken by the spacecraft Giotto during its March 13 1986 encounter with the comet.

From Earth, we see the ejected material pushed by the pressure of sunlight to form a long tail like this:

Comets are relatively short-lived objects that eventually break up after many close approaches to the Sun. The dust clouds from these break-ups cause meteor showers as the Earth passes through them. New comets are sent into the inner solar system by encounters with the planets. Here is the sequence of events that sent the new comet Wild-2 on its way to solar encounter near the Earth:

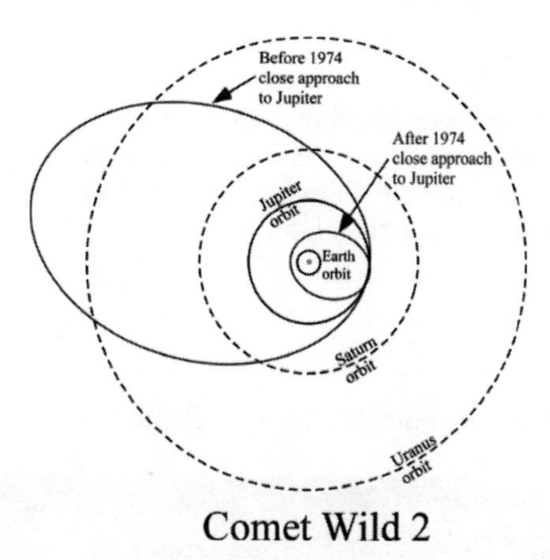

Comet Wild 2

On January 2, 2004, this relatively new comet was met by the Stardust spacecraft which took close-up pictures of it

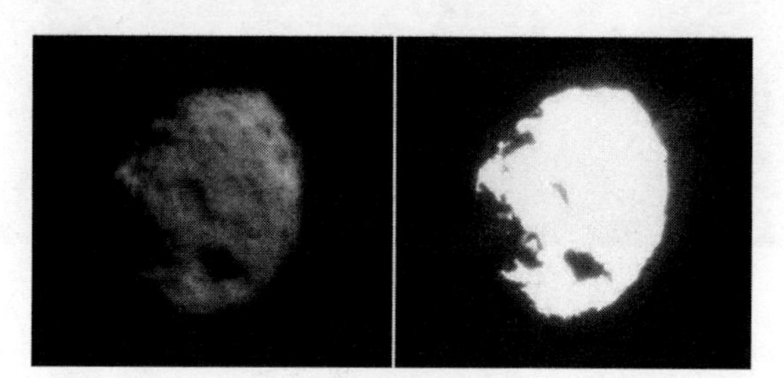

and captured some of its dust. The dust samples were safely returned to Earth for analysis on January 15, 2006.

010.6 The Kuiper Belt

The region beyond Neptune is now known to have something like 100,000 comet-like objects. Like comets, they are made of frozen gas and ice with some dust mixed in. In fact, they become comets when close encounters with Neptune send them into the inner solar system. This region is called the *Kuiper Belt* after Gerard Kuiper who predicted that the gravitational effects of the planet Pluto would prevent such a belt from lasting until the present time. (That's right They named it after the fellow who predicted that it would *not* be found. Falsificationism at its best!)

Alas poor Pluto!

Pluto and its moon Charon, shown here in a Hubble Space Telescope picture

are typical Kuiper Belt objects. Each of them is made almost entirely of ice. Until recently Pluto could at least claim to be the largest of the Kuiper Belt objects. However that is no longer true. Eris, with a satellite Dysnomia

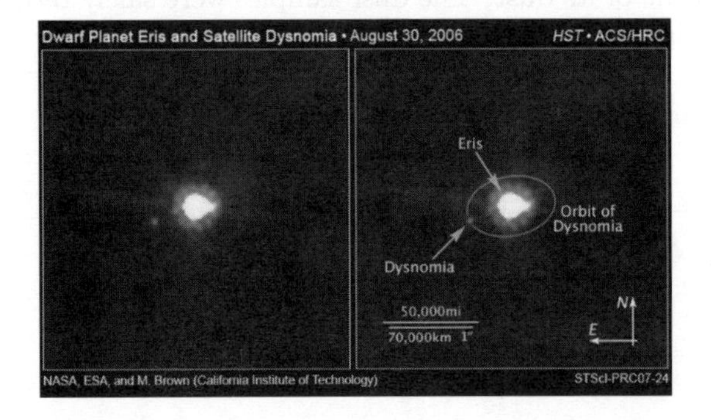

now takes the honor of being the largest Kuiper Belt Object. Here is the orbit of Eris (formerly known as 2003 UB313 or, in some circles, as 'Xena') and some size comparisons:

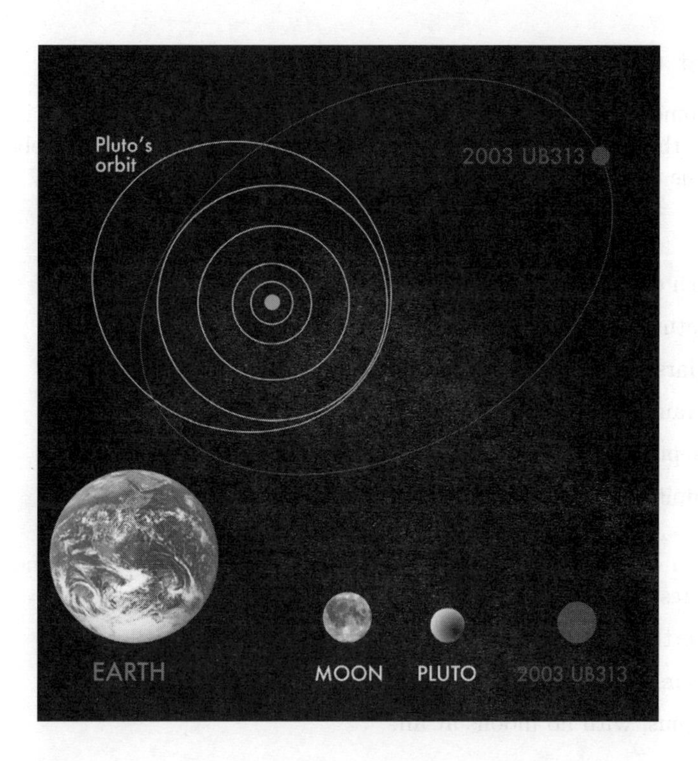

A Crowded Sky

Now that the technology to discover Kuiper Belt Objects exists, they are being discovered at a very rapid rate. An up-to-date map of all objects found so far can be found through a link on our website to the Minor Planets Centre.

010 Spot Check

Here are some questions to check your understanding of the material in module 010. Both the answers and where to find these questions at the website may found at the end of the Study Guide.

1 Which of these planets is the farthest from the Sun?

a. Saturn

b. Mars

c. Uranus

d. Neptune

e. Jupiter

2 Which of the following three systems is regarded as the most normal for a terrestrial planet?

a. Earth, with a moon larger than the dwarf planet Pluto.

b. Mars, with two moons each the size of an asteroid.

c. Venus, with no moons at all.

3 Asteroids are usually made of

a. Styrofoam and possibly poster paint.

b. rock and possibly iron.

c. ice and possibly frozen gas.

d. concrete and possibly marble.

e. gold and possibly silver.

4 A planet with a large system of moons would have to be a

a. Jovian Planet.

b. Kuiper Belt object.

c. terrestrial planet.

5 Pluto is now regarded as

a. the largest dwarf planet in the Kuiper Belt.

b. the smallest dwarf planet in the Kuiper Belt.

c. one of the larger dwarf planets in the Kuiper Belt.

d. a planet that happens to be in the Kuiper Belt.

6 Which of the following objects would be most likely to have a long elliptical orbit that takes it from far outside the orbit of Mars to a close approach to the Sun?

a. planet.

b. asteroid.

c. comet.

011: The Terrestrial Planets

011.1 Mercury

Weather and Surface Conditions

Mercury has almost no atmosphere and rotates very slowly, resulting in an extreme temperature range.

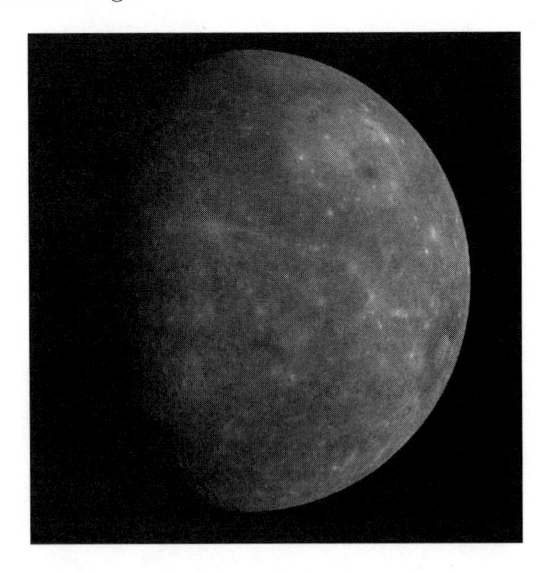

At noon, temperatures reach 400 C. During the three month night, the temperature drops to −150C. There are no volcanos or earthquakes.

Mercury has an axial tilt of only 1/100 of a degree, so there are no seasons and the bottoms of craters near the poles are never lit by the Sun. With no atmosphere to transfer heat from warmer regions, it can stay very cold at the poles and it is thought that large amounts of water ice may exist there. Radar reflections from the polar regions are consistent with ice.

Orbit and Rotation

Mercury has the most elongated (eccentric) orbit of all the planets. Its distance from the Sun varies from 46 million kilometers to 70 million kilometers. Its rotation on its axis is extremely slow and that, combined with the way that the planet speeds up its motion at perihelion (in accord with Kepler's Equal Area Law) gives rise to an odd effect: The Sun exhibits retrograde loops. It can rise, stop in the sky and then reverse to set in the same place it rose and then rise again from the same spot before proceeding across to the other horizon.

It takes 88 days to complete one orbit and it takes 59 days to complete one rotation relative to the fixed stars. That turns out to be a 3:2 resonance as you can see in this picture:

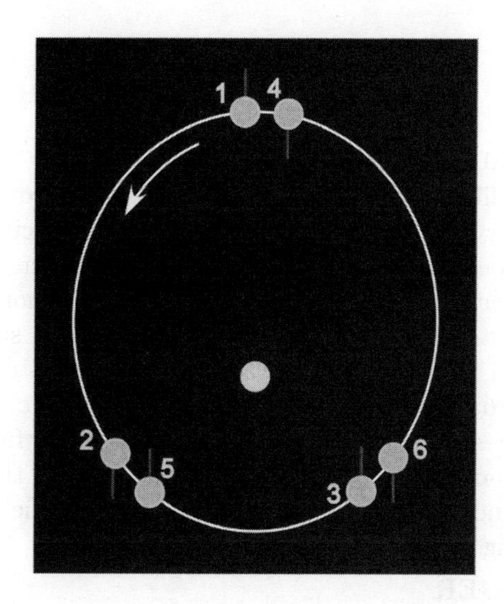

During one orbit, Mercury complete 1.5 rotations, so someone who starts the 'year' at solar noon will end it at exactly midnight. A second orbit brings the planets orientation back to where it started after 3 full rotations on its axis.

The fact that Mercury gets so close to the Sun provided one of the first tests of *Einstein's Theory of Gravitation.* In Einstein's Theory, gravity is actually due to the curvature of space and time. His theory predicts that space close to the Sun will be curved by its gravity. One effect of that curvature is that there are no longer 360 degrees in a circle around the Sun. At the orbit of Mercury, there is a missing angle of about a tenth of a second of arc. Each time Mercury orbits the Sun, the long axis of its elliptical orbit shifts by that missing half second of arc. Over the course of 100 years, the shift adds up to about 43 seconds of arc. The shift was noticed and had been puzzling astronomers for some time when Einstein's theory came along to explain it. The effect is called the *relativistic perihelion precession.*

Magnetic Field

One surprising result from probes of Mercury is that it does have a small magnetic field, with about 1% of the intensity of the Earth's field. That was a surprise because our current model of how planets acquire magnetic fields requires rotating currents in a liquid metal core. The rotation rate of Mercury was thought to be too small for that process to happen.

The relatively high density of Mercury suggests that it has an iron core that is much larger in proportion to its size than that of any other planet. That may explain how it can generate a magnetic field with such a slow rotation rate.

Moons

Mercury has no natural moons of any kind.

Space Probes

1974 Mariner 10

Mercury is very difficult to get to from the Earth. A direct transfer orbit from Earth to Mercury requires very large changes in velocity at each end, far more than existing rockets are capable of. Mariner 10, launched in 1974, went first to Venus and then used the gravity of Venus to slingshot it on to Mercury. This *gravitational slingshot* maneuver has since become a standard technique but was brand new in 1974. The spacecraft ended up in its own eccentric orbit around the Sun, approaching Mercury on every other orbit of Mercury. It made three close approaches before it ran out of maneuvering fuel. During those passes it mapped about half of the surface and discovered that Mercury has a surprising large magnetic field (about 1% of the Earth's field). Mariner 10 is still out there, passing Mercury every other orbit.

2004 MESSENGER

The second mission to Mercury has only just now arrived at the planet. It is called MESSENGER. The picture at the top of this section was taken by that spacecraft. MESSENGER was launched August 3, 2004 and made several gravitational slingshot passes to adjust its orbit. It flew past the Earth in August 2005 to get a boost from its gravity and then past Venus twice, once in October 2006 and again in June 2007. Its first Mercury pass was January 14, 2008. It will have two more passes in October 2008 and September 2009. It will then go into orbit around Mercury and begin a full year mapping mission. The probe will be measuring Mercury's magnetic and gravitational fields to learn more about the interior of the planet. It will also be looking much more closely for ice at the poles.

A third mission to Mercury is planned by the European Space Agency in a joint mission with Japan to be launched using a Russian Soyuz rocket in 2013. The mission is called BepiColumbo and will consist of two different probes, one to be placed in an elliptical orbit around Mercury and one to be placed in a circular orbit. In addition to reproducing the observations of MESSENGER with a different set of instruments, the probe will also be carrying out new tests of Einstein's Theory of Gravity.

011.2 Venus

Weather and Surface Conditions

Venus has a dense atmosphere, about 90 times the atmospheric pressure that we have on Earth. The atmosphere consists mostly of Carbon Dioxide, giving the planet a severe case of global warming. Temperatures reach 450 C, making it the hottest planet in the Solar System. The clouds that block our view of the surface are mostly made of sulfuric acid and 400 mile per hour winds are common among the cloud tops but there is very little wind at the surface.

Just to complete the picture, there is almost no water at all and there are active volcanoes on the surface. The dense atmosphere bends light toward the surface, so if you were standing on Venus, you would always have to look up a bit to see the horizon. In other words, we have temperatures hot enough to melt lead, fire and brimstone, sulfuric acid clouds, and wherever you stand it looks as if you are standing in a hole. The surface of Venus is possibly the most hostile to life of any in the Solar System.

Orbit and Rotation

Venus has the most nearly circular orbit of any of the planets and orbits the Sun once every 224 Earth-days. It rotates *backward* (clockwise if viewed from far above the Earth's North Pole) once every 243 days. On this planet, the Sun rises in the West and sets in the East.

The orbit and rotation of Venus display an odd synchronization with the Earth. Each time that Venus has a closest approach to Earth, the same side is always facing us. The origin and significance of this synchronization are not known.

Magnetic Field

Venus has no magnetic field at all. A magnetic field is usually generated by rotation and a liquid metal core. Evidently its rotation rate is just too slow and its iron core may be mostly solid.

Moons

Venus has no natural moons of any kind.

Space Probes

Venus has been visited by about 20 spacecraft (out of 41 attempts), starting with Mariner 2 in 1962. The earliest probes established the very dense atmosphere of carbon dioxide and the high surface temperature. Just a few of the probes are mentioned here.

1970: Venera 7

was the first spacecraft to actually land on another planet. It was the 16th Russian attempt to get a probe to Venus. A small capsule entered the atmosphere and descended by parachute. Weak radio signals were received from the probe during its 35 minute descent and for 23 minutes after landing.

1975: Venera 9

In addition to its heat shield for atmospheric entry, Venera 9 had a cooling system that permitted it to operate on the surface for about 53 minutes. It found the surface pressure to be about 90 times the pressure at the surface of the Earth and the light-level was about like Earth on a cloudy summer day. Television pictures of the surface were sent back: Venera 9 was followed by Venera 10, 11, 12, 13, 14, 15, 16. The Venera sequence ended in 1983 with the radar mapping orbiters Venera 15 and 16 and was followed by two more landers, Vega 1 and Vega 2 in 1984.

1978: Pioneer Venus Orbiter

Launched in May 20, 1978 and inserted into orbit around Venus on December 4, 1978. Radar mapping of the planet was done until July 1980. In 1991 and 1992 the Radar Mapper was reactivated to finish mapping parts of the planet that had not been accessible before. Here is a topographic map of the surface produced by the orbiter:

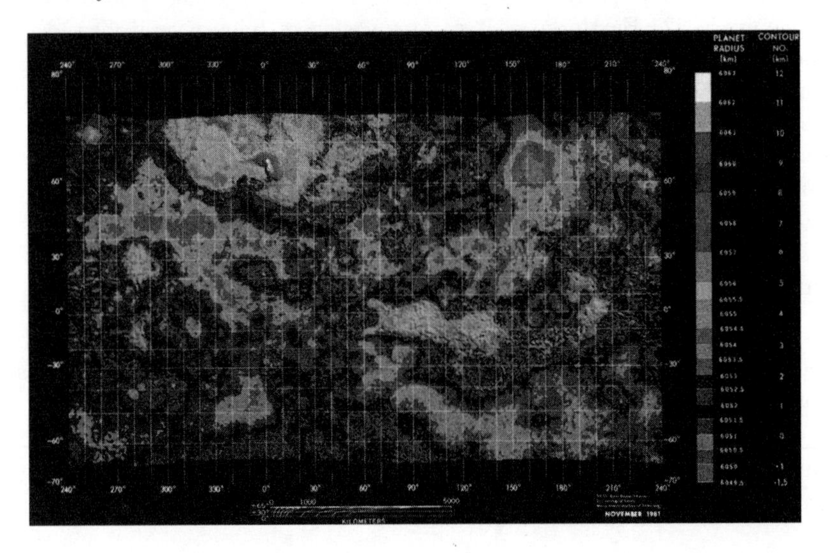

The orbiter also carried infrared and ultraviolet cameras that revealed information about the upper atmosphere of the planet. Here is an ultraviolet image that shows the cloud structure quite clearly.

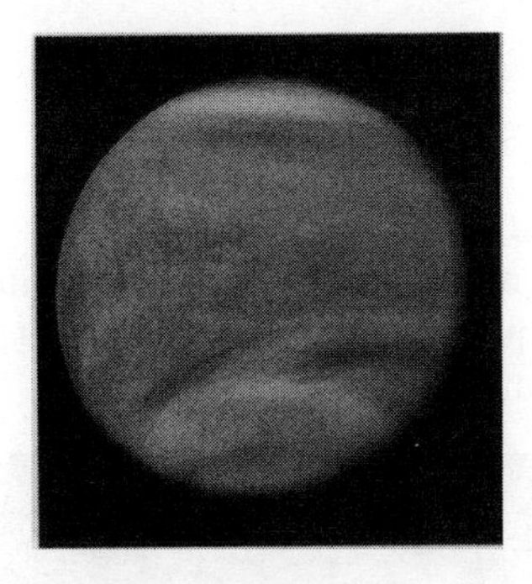

1978: Pioneer Venus Multiprobe

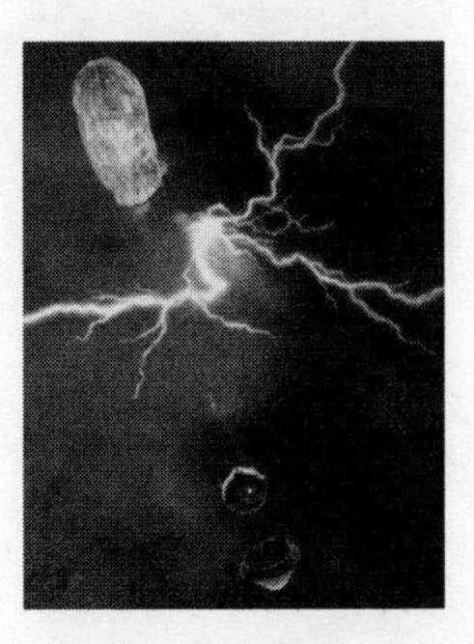

The Multiprobe consisted of a 'bus' and four atmospheric probes. All four probes entered the atmosphere of Venus on Dec. 9, 1978, followed by the bus. The largest probe is the one shown in this artist's rendering. It consisted of a spherical pressure vessel protected by a heat shield for atmospheric entry and a parachute. Three smaller probes were also in heat-shielded spherical pressure vessels but without parachutes. Each probe sent back radio signals throughout its descent, reporting on atmospheric conditions. Of the four probes, only the small one sent to the day side of the planet survived the descent. That probe broadcast from the surface for about an hour. The spacecraft bus also carried instruments but no heat shield and provided information about the upper atmosphere of the planet before it burned up.

1989: Magellan

Launched from the Space Shuttle on May 4, 1989, Magellan entered a polar orbit around Venus on August 10, 1990 and produced detailed, three-dimensional maps of the surface with a resolution of about 100 meters.

2005: Venus Express Launched from Baikonur Cosmodrome, November 9, 2005, this European Space Agency probe was inserted into Venus orbit on April 11, 2006 and has been sending back data on the atmosphere of Venus since then.

011.3 Earth

Weather and Surface Conditions

Earth is the one planet whose temperature ranges include the *triple point of water* with a pressure far enough above the triple point pressure to have water in all three forms, solid, liquid and gas, at the same time. You can plainly see all three forms in this picture.

Only here can you have oceans, icebergs, clouds, rain, and snow.

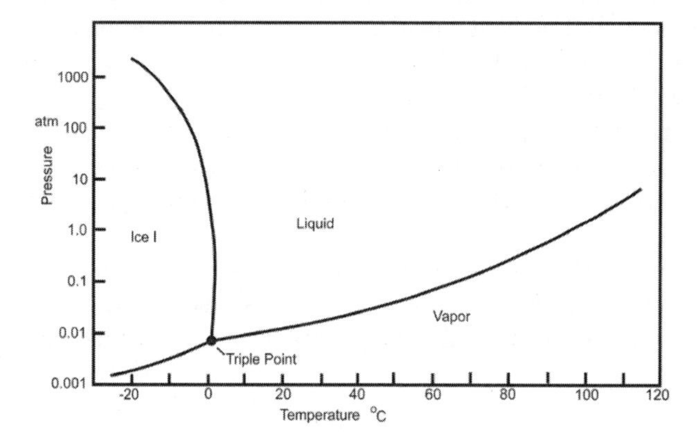

One fifth of the atmosphere consists of one of the most reactive and corrosive elements known, Oxygen. The surface is the least stable of any of the planets, consisting of many free-floating plates that cause volcanoes, earthquakes, and active mountain building as they move around and bump into each other. It also has a very large Moon that is very difficult to explain.

Orbit and Rotation

The Earth rotates on its axis once every 24 hours (minus about 4 minutes) and has a rotation axis that is inclined at 23.5 degrees from the perpendicular to the plane of its orbit. For the northern hemisphere, the rotation axis is inclined directly away from the Sun every December 21 or 22. That is the Winter Solstice. The axis is inclined directly toward the Sun, the Summer Solstice, every June 20 or 21. As we discussed earlier, that axial tilt is responsible for the seasons.

The Earth's orbit is not quite circular. We are at closest approach to the Sun, or perihelion, every January 3. The largest distance from the Sun, or aphelion occurs every July 4. The perihelion distance is 147,300,000 kilometers while the aphelion distance is 152,100,000 kilometers. That is enough to make about a 6% difference in the intensity of sunlight hitting the Earth. Notice that, for the northern hemisphere, the closest approach to the Sun occurs in the winter so that the difference actually works *against* the seasons.

The Earth's rotation axis is known to precess and wobble, and the long axis of the Earth's orbit also moves around over long periods of time. In addition, the eccentricity of the Earth's orbit varies so that the difference in sunlight intensity between perihelion and aphelion can be as large as 23%. The result is a cyclical shifting of the relationship between the seasons and the intensity of sunlight and is the basis for the idea of Milankovich cycles to account for long-term climate changes. Unfortunately the idea does not seem to work very well. The actual climate of the Earth is evidently too complex to respond simply to changes in solar input.

Magnetic Field

The Earth has an extensive magnetic field that plays an essential role in protecting the planet from radiation. Charged particles streaming from the Sun are trapped by the magnetic field in regions called the Van Allen Belts and enter the atmosphere only near the North and South poles, where they form the Aurora Borealis and the Aurora Australis. The magnetic field is thought to result from the Earth's rapid rotation rate and the presence of a large amount of liquid iron in the core.

Moons

Earth has one natural moon and about 8000 artificial ones. The Moon is about 1/4 the size of the Earth. That relative size is very unusual and is one of several indications that our Moon did not form along with the Earth.

Space Probes

There are currently about 8000 man-made objects in orbit around the Earth. About 500 of them are currently operational satellites. A large number of them are Earth-observing satellites that are responsible for much of what we know about our own planet.

011.4 Earth's Moon

Surface Features

Large, circular, dark areas make up the "face" of the "Man in the Moon." These areas are made of basalt, the same material as the Moon's mantle and are lower and younger than the rest of the Moon's surface.

They are probably the result of ancient impacts that caused liquid mantle to flow out over the Moon's surface. Almost all of the maria are on the side of the Moon that faces the Earth.

Each crater shows at least a ring wall where the crust has been pushed back and folded over by the impact. Often there is a central peak where the surface rebounded from the impact and radial streaks left by ejecta - material thrown out across the surrounding surface.

The Crater Copernicus is 90 kilometers in diameter. Many smaller craters can be seen near it.

Orbit and Rotation

The Moon orbits the Earth once every 27 days (relative to the distant stars) and rotates on its axis with exactly the same period so that it always has the same side facing the Earth.

This situation is due to the Earth's gravity raising bulges on the Moon's surface which have slowed its rotation relative to the Earth until it is now in what is called 'tidal lock.'

The side that always faces away from the Earth is actually quite different from the side that we can see, having very few maria and a much thicker crust.

The Moon's orbit is significantly elliptical. A typical perigee distance is 360,000 kilometers while a typical apogee distance is 400,000 kilometers. One result is that a solar eclipse when the moon is at its apogee cannot be total because the apparent size of the Moon is smaller than the apparent size of the Sun. Instead of a total eclipse, we get an annular eclipse in that situation.

Another consequence of the elliptical orbit is that the Moon's rotation on its axis does not exactly match its position along its orbit and it seems to wobble back and forth, showing us a bit of its far side.

Magnetic Field

The Moon has no magnetic field of its own. That observation is consistent with the current model of how planets form magnetic fields since the Moon rotates slowly and its low density suggests that it does not have an iron core.

Oddities of the Moon

The Moon is 1/4 the size of the Earth. The largest satellite of Saturn is about 1/25 the size of Saturn.

The Moon's orbit is tilted relative to the Earth's equator by an amount that wobbles a bit but is never less than 18.5°.

The low density of the Moon implies an almost complete lack of iron.

The surface material from the Moon has a chemical composition identical to that of Earth's surface with one exception: It is extremely dry.

The dryness of moon rock does not just refer to the absence of ordinary water. Earth rocks contain large amounts of water that is chemically bound in the form of extremely stable hydrates. Hydrates are missing from Moon rocks.

The Formation of the Moon

We cannot really discuss the formation of the Moon until we set the scene with a model for the formation of the entire solar system. However, we need just one detail of the currently accepted scenario: Water could not have condensed out near the hot new sun along with the rocks and iron of the inner planets and could only have come from a rain of frozen comets that originally formed farther out.

With the scene properly set, here are the observational constraints on any model of how the Moon formed:

The tilt of the Moon's orbit argues that its formation involved an outside body.

The similarities in chemical composition imply that all of the objects involved formed in the neighborhood of Earth's orbit.

The lack of iron on the Moon implies that the Moon could not be a captured object that formed separately.

The dryness of lunar rocks implies that they were heated to high temperatures after the rain of comets that supplied water to most terrestrial objects.

The currently accepted theory is the Collision Model, sometimes called the Big Splash Theory. It is thought that shortly after the formation of the Earth and the initial rain of comets that supplied water and gas, an object the size of Mars collided with the proto-Earth.

Both planets were re-melted by the collision and the iron core of the intruder sank into the Earth while lighter mantle material was blown off in a giant splash to form a ring.

The Moon formed from mantle material that had been remelted by the collision, which explains how the water was driven out of the hydrated minerals.

Space Probes

There have been about 72 attempted lunar missions with about 48 successes between 1958 and 1998. The first successful soft landing on the Moon was Surveyor 1, in 1966. A series of Surveyor landings were made in 1967 and 1968 and were followed by the Apollo program.

Apollo 8: Dec. 1968 Orbited the Moon (10 orbits)

Apollo 10: May 1969 Practiced docking maneuvers in lunar orbit.

Apollo 11: July 1969 First humans on the Moon, returned 46 pounds of Moon rocks to Earth.

Apollo 12 Nov. 1969 Retrieved pieces of Surveyor 3

Apollo 13 Apr. 1970 Abort and return

Apollo 14 Jan/Feb 1971

Apollo 15 Jul/Aug 1971 First use of lunar roving vehicle

Apollo 16 Apr. 1972

Apollo 17 Dec 1972 Last human mission to the Moon The Apollo series of missions returned 842 pounds of Moon rocks for study on Earth.

The U.S.S.R. launched a long series of Moon missions, ending with LUNA 21 in 1973 which landed a robotic lunar rover, and LUNA 24 in 1976 which returned lunar samples to Earth.

After a considerable hiatus, space probes have begun to return to the Moon:

Hiten (Muses-A) Japan Jan 1990 Lunar orbiter

Clementine, USA, Jan 1994 Lunar orbiter

Lunar Prospector, USA, Jan 1998 Lunar orbiter searched for water ice near the Moon's poles.

SMART-1, ESA (European Space Agency), Launched Sep 2003, impacted on Moon Sep 2006. This probe tested an ion propulsion system.

KAGUYA (SELENE) JAXA (Japan Space Agency), Launched Sep 2007, Entered Lunar Orbit Oct. 18, 2007, One main orbiter and two smaller ones. Shifted to regular observation mode on December 21, 2007. Will orbit the Moon for ten months.

Chang'e 1 China, Launched October 24, 2007, lunar orbiter, returning data now.

Chandrayaan-1, ISRO (Indian Space Research Organisation) to launch in April 2008, lunar orbiter.

Lunar Reconnaissance Orbiter, NASA, to launch in Fall, 2008, lunar orbiter

011.5 Mars

Weather and Surface Conditions

Mars has an atmosphere of mostly carbon dioxide with about 1% of the Earth's atmospheric pressure. It is mostly very cold with temperatures quite similar to those in Antarctica here on Earth. The current temperature records for Antarctica (high of 59F, low of −129F) are essentially the same as we would expect on Mars. Water exists only as water vapor in the thin atmosphere and as ice in the polar ice caps and possibly underground as a permafrost layer. However there are abundant surface features that suggest the presence of liquid water in the past and a few indications, as shown below, that water still appears on the surface even now.

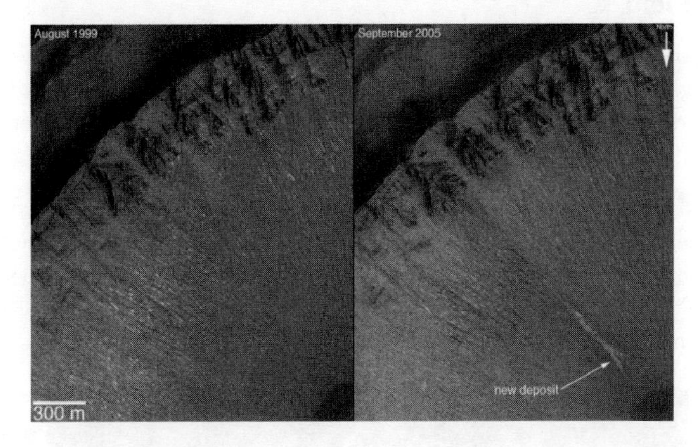

Although the atmosphere is thin, it is enough to form clouds and duststorms.

Orbit and Rotation

A Martian *sol* of 24 hours and 39 minutes is very similar to Earth's 24 hour solar day and its rotation axis is tilted in a similar way (at 25.2 degrees from the perpendicular to its orbit), giving rise to the same sort of seasons that we have here.

Mars orbits the Sun once every 687 days. Its orbit is very far from circular with a perihelion distance of 128.6 million miles and an aphelion distance of 160 million miles. The intensity of sunlight changes by a drastic 40% between these extremes. Just as for the Earth, the rotation axis and the long axis of the elliptical orbit both precess over long periods of time, so there should be Milankovich cycles that are much more dominant than on Earth. Speculations about drastic shifts in Martian climate and even in atmospheric pressure are sometimes based on these cycles. There is also evidence from NASA's Mars Global Surveyor and Mars Odyssey missions that Mars is now coming out of an ice age and looked like this within the past million years:

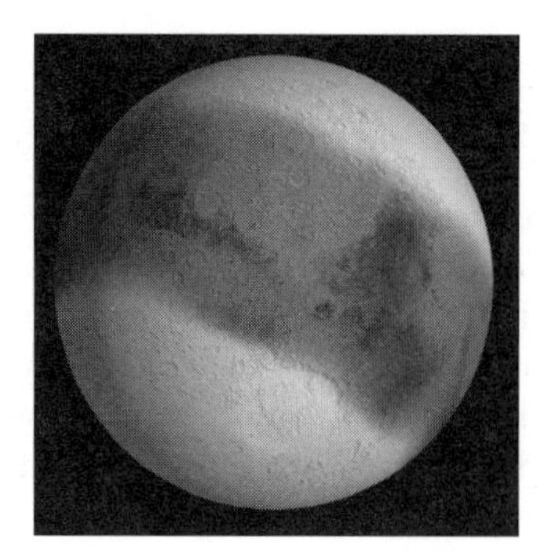

Magnetic Field

Mars does not appear to have an overall magnetic field of its own. However there are small regions of magnetic field that are thought to be ore deposits that were magnetized during an earlier time when the planet did have a magnetic field.

Moons

Mars has two small moons, Phobos and Deimos. In the picture below, Phobos is at the lower right while Deimos is at the lower left, both imaged by the Viking 1 Orbiter in 1977. The object at the top is the asteroid Gaspra imaged by the Galileo spacecraft in 1991.

Phobos, the larger of the two moons is about 17 miles across and, as can be seen in this 2008 Mars Reconnaissance Orbiter image, has a six-mile wide crater (Stickney Crater).

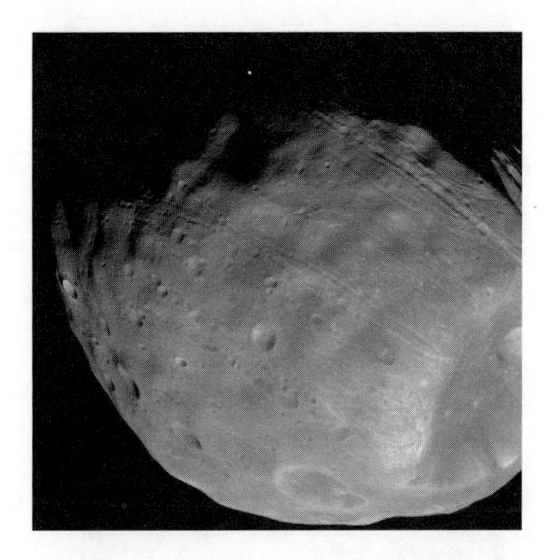

Phobos completes an orbit of Mars in 7 hours and 39 minutes. Since that is faster than the planet rotates, Phobos rises in the West and sets in the East. Its orbit has a periapsis of 9235.6 kilometers or 2.76 times the radius of Mars, which brings it closer to the surface of its primary than any other moon in the Solar System. For comparison, the orbit of Deimos has a periapsis about 7 times the radius of mars.

The moons were not discovered until 1877 although Jonathan Swift wrote about them in Gulliver's Travels in 1726. However, they are easy to see from Mars, as this sequence of pictures taken by the Spirit Mars rover shows

Because the time intervals were shorter for the first three pictures, you can see that the two Moons are going in opposite directions. Deimos, with a period of 30 hours and 18 minutes, orbits slowly enough to rise in the East and set in the West like a normal moon.

Space Probes

There have been 36 space probes launched toward Mars with about 15 successful missions. Here are a few of them

1964: Mariner 4 (flyby) was the first mission to successfully send back pictures, including this one

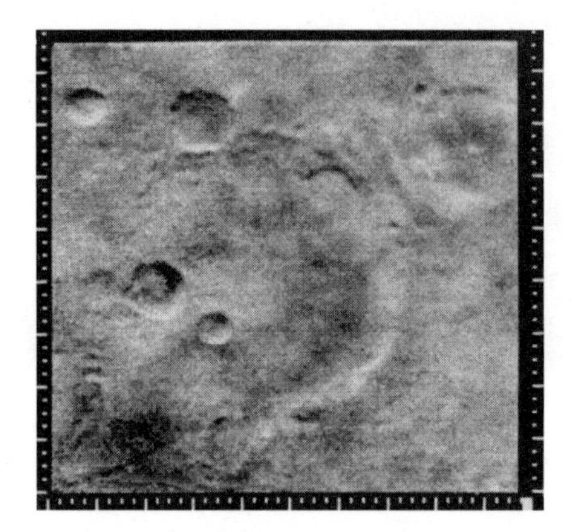

which includes the 151 kilometer diameter Mariner Crater.

1971: Mariner 9 (orbiter) Launched on May 30, 1971. Arrived in Mars orbit on November 14, 1971. When it arrived, its view of the surface was completely obscured by a planet-wide dust storm. By January, 1972, the dust had settled and the orbiter was able to map the surface with many images such as this one of the volcano Olympus Mons standing above the dust storm

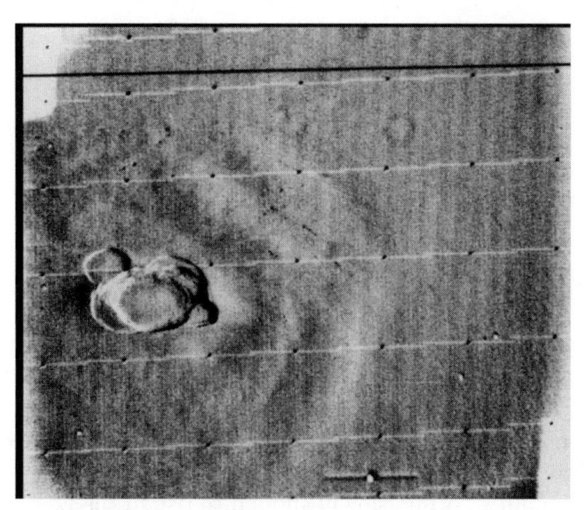

1975: Viking 1 and 2 (landers) Launched on August 20 and September 9, 1975. Arrived in Mars orbit on June 19 and August 7, 1976. Landed July 20 and September 3, 1976. Viking 1 landed on Chryse Planitia while Viking 2 landed at Utopia Planitia. Viking Orbiter 1 spent 4 years in Mars orbit. The landers were powered by radioisotope thermal generators which lasted for years. Viking Lander 2 went silent on April 11, 1980. Viking Lander 1 transmitted images and other data until November 11, 1982.

These landers were designed to scoop up some Martian soil for analysis as part of the search for life. Here you can see the scoop in the lower center of this picture from Viking 1.

Viking 1 Landing Site

Viking 1 on Mars

This area was named the "Sandy Flats" area at Chryse Planitia. The image of the landing area was taken by Viking Orbiter 1.

The landing site of Viking 2, in the northern Utopia Planitia, was quite different looking.

with lots of large boulders everywhere. A composite of the images produced by both orbiters gives a striking image of Mars

Schiaparelli Hemisphere

Syrtis Major Planus Region

1996 Mars Global Surveyor Launched November 7, 1996. Entered Mars orbit using aerobraking on Sept 11, 1997.

The orbiter remained active until November 22, 2006. During its ten year mission, it took 240,000 pictures of the planet surface, used its spectrometer to discover concentrations of hematite, a mineral that normally forms only with water, and found remnant magnetic fields indicating that Mars once had a magnetic field similar to Earth's. The hematite-rich region that it found was selected as the landing site for the rover Opportunity.

1996 Pathfinder Launched on December 4, 1996 and landed on Mars on July 4, 1997 using an airbag system instead of a fully controlled hover type of descent. Here is a composite of pictures taken from the camera on the lander as the Sojourner rover went about its business

Its last transmission from Mars was on September 27, 1997. Intended as a sort of toy testbed for ideas that would be used in later missions, it succeeded far beyond expectations.

2003 Spirit and Opportunity Launched June 10, and July 7 2003 and landed January 3 and January 24 2004. Landing was by parachute and airbag just as pathfinder had demonstrated and, at each landing site, a grown-up version of the sojourner rover rolled onto the surface of Mars. Here is what the Spirit rover saw from the top of a low plateau where it spent the last months of 2007

Spirit has travelled about 4.5 miles so far and Opportunity has gone about 7.2 miles. They are expected to last through 2009.

011 Spot Check

Here are some questions to check your understanding of the material in module 011. Both the answers and where to find these questions at the website may found at the end of the Study Guide.

1 Viking 1 and 2 were sent to explore

a. Earth's Moon.

b. the planet Jupiter.

c. the planet Neptune.

d. the planet Venus.

e. the planet Mars.

2 The Moon's orbit around the Earth

a. is exactly circular.

b. is elliptical enough to give us an annular lunar eclipse when the Moon is near its apogee.

c. is somewhat elliptical but not enough to affect eclipses.

d. is elliptical enough to give us an annular solar eclipse when the Moon is near its apogee.

3 On the surface of Mars, water exists

a. in all three forms, solid, liquid, and gas.

b. Only as ice.

c. Only as water vapor.

d. Mostly as ice and water vapor.

4 The statement that lunar material is much "dryer" than Earth material refers to the absence of

a. liquid water.

b. hydrated minerals.

c. ice.

d. mud.

5 The presence of frozen water on Venus is

a. possible because it has no axial tilt so that its poles never face the Sun.

b. impossible because every part of it is too hot for water ice.

c. impossible because the night side, where it is cold, eventually rotates to face the Sun.

d. possible because the planet always keeps the same side turned away from the Sun.

6 So far (as of 2008), the planet Venus has been visited by

a. just one successful space probe.

b. about 20 successful space probes.

c. two or three successful space probes.

d. no successful space probes.

e. about 48 successful space probes.

7 Which of the following planets has just two moons?

a. Earth

b. Neptune

c. Jupiter

d. Venus

e. Mars

8 The first space probe to reach the planet Mercury was

a. Pioneer 10.

b. Mariner 10.

c. Mariner 9.

d. Pioneer 11.

e. MESSENGER.

9 Venus

a. always keeps the same side toward the Sun so that solar time never changes.

b. has a solar day that last for three complete orbits around the Sun.

c. rotates backwards so that the Sun rises in the West.

d. has a solar day that is very close to an Earth day in length.

e. has a solar day that last for two complete orbits around the Sun.

10 Mars' orbit is currently

a. elliptical enough to make the intensity of sunlight vary by 6 percent.

b. slightly elliptical but not enough to affect the intensity of sunlight.

c. exactly circular.

d. elliptical enough to make the intensity of sunlight vary by 40 percent.

11 Ancient lava flows on the Moon are called Lunar

a. planitia.

b. valleys.

c. maria.

d. terrae.

e. craters.

12 Earth's orbit is currently

a. elliptical enough to make the intensity of sunlight vary by 40 percent.

b. exactly circular.

c. elliptical enough to make the intensity of sunlight vary by 6 percent.

d. slightly elliptical but not enough to affect the intensity of sunlight.

13 The first human landing on the Moon was Apollo 11 in

a. 1976.

b. 1982.

c. 1966.

d. 1969.

e. 1972.

14 Which of the following planets can be said to have almost no atmosphere?

a. Venus

b. Mercury

c. Mars

d. Earth

012: The Jovian Planets

012.1 Jupiter

0.0.1 Surface

Cassini-Huygens Image: NASA/JPL

Jupiter is the largest of the Gas-giant planets. It does not have a solid surface and is basically a large ball of mostly hydrogen and helium gas with smaller amounts of other elements. Current models of its interior suggest that the gas density and pressure increase going toward the center until the hydrogen is compressed to a liquid metallic state. The surface that can be seen in the true-color mosaic shown here (from the Cassini flyby on Dec. 29, 2000) consists of clouds of ammonia, hydrogen sulfide, and water. The bands are circulating in opposite directions, causing turbulent vortexes to form between them. The largest of these is called the Great Red Spot.

At 11 times the size of the Earth, Jupiter is thought to be about as large as a planet can get. Current models of its interior indicate that adding more material to it would actually cause it to become more compressed so that it would shrink. Here are some size comparisons to the Earth, and the SunNotice that the Great Red Spot is actually quite a bit larger than the Earth.

Because it is more than five times as far from the Sun as the Earth, the intensity of sunlight at Jupiter is about 1/25 of what it is here. In fact, the surface of Jupiter receives about as much heat from the interior of Jupiter as it does from the Sun. The effective surface temperature is 125 Kelvin or −148C or 234 Fahrenheit degrees below zero.

Orbit and Rotation

Jupiter rotates on its axis once every 10 hours. Actually it does not rotate like a solid ball and rotates a bit faster at the equator. It takes 9 hours and 56 minutes for its surface to rotate once near the poles and 9 hours and 50 minutes near the equator. The rotation rate is so rapid that the planet bulges noticeably at the equator. Its rotation axis is almost perpendicular to the plane of its orbit (3.13 degrees from the perpendicular) so that there are no real seasons.

Its orbit is elliptical with a perihelion distance of 4.952 au and an aphelion distance of 5.455 au. and takes it around the Sun every 4330 Earth days (11.86 years).

0.0.2 Magnetic Field

Jupiter has a magnetic field with about ten times the intensity of the Earth's field over a much larger region of space. The currently accepted model of how planets generate magnetic fields requires rotation and a liquid metallic conductor at the planet's core. Jupiter is certainly rotating fast enough but it is not dense enough to have an iron core. It is thought that the extreme pressure at the center of the planet causes hydrogen to become a liquid metal. Until 1996, when it was finally produced in the laboratory (see Ladbury, R. "Livermore's Big Guns Produce Liquid Metallic Hydrogen." *Physics Today* **49**, 17–18, May 1996), this liquid metallic phase of hydrogen was purely theoretical and the magnetic field of Jupiter was the only observational evidence that it existed.

Just as is the case for Earth, the magnetic North and South poles are displaced from the poles defined by the planets rotation axis. That causes the magnetic poles to be carried with the planet's rotation and provides a reliable way to measure the rotation of this non-solid planet.

Moons and Rings:

Jupiter has a large system of Moons, all orbiting close to the plane of its equator. Jupiter has been described as a "failed star" because its composition is quite similar to a typical star like our Sun. It just did not grow large enough to ignite nuclear fusion reactions at its core. However it has formed its own miniature solar system. More moons are being discovered, so the number of Moons keeps changing. As of this writing, there are 62. The most recent ones only have catalog numbers like 2003 J10.

The largest moons were first noted by Galileo and are called the *Galilean Satellites of Jupiter.*

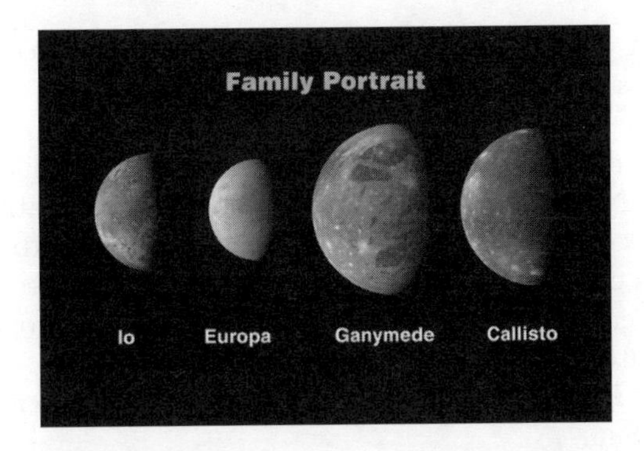

Each of these moons is a world in its own right. We would not hesitate to describe these objects as planets if they were in orbit around the Sun. Ganymede, the largest of the moons of Jupiter is larger than the planet Mercury.

We will be discussing these moons in detail later. Two of them, Europa and Ganymede, appear to have oceans of liquid water beneath their surfaces and are high on the list of places to look for life.

Jupiter, like all of the other gas giant planets, has a system of rings. They are made of dark colored rocks and are hard to see.

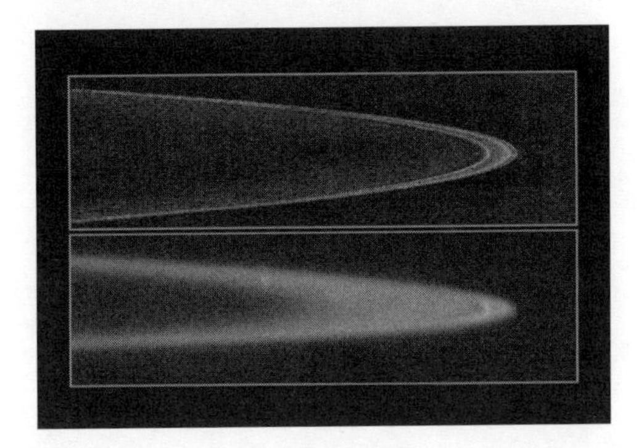

These pictures were taken by the New Horizons probe as it sped by on its way to Pluto. The top picture is taken looking away from the Sun as the probe approached Jupiter. The two thin rings are clearly visible. The bottom picture was taken looking back toward the Sun as the probe sped away and shows the haze of dust surrounding the main rings. Small moons, called "shepherd moons" confine the rings through gravitational interactions.

Space Probes

1972 Pioneer 10 NASA Flew past Jupiter at a distance of 81,000 miles from the planet on Dec. 3, 1973. Analyzed the composition of Jupiter's atmosphere and discovered Jupiter's enormous magnetic field and radiation belt and is still heading out of the solar system.

1973 Pioneer 11 NASA Flew past Jupiter at a distance of 27,000 miles in Dec. 1974 on its way to Saturn.

1977 Voyager 1, Voyager 2 NASA arrived at Jupiter on March 5, 1979 and July 9, 1979 and made the first detailed maps of the Galilean satellites.

1989 Galileo NASA Arrived at Jupiter and released an atmospheric probe in July 1995. In December 1995 the probe penetrated deep into the cloud layers and measured the composition of the atmosphere while Galileo entered Jupiter orbit. Galileo operated until 2003 and discovered that Europa may have an ocean of liquid water below its frozen crust.

1990 Ulysses ESA flew past Jupiter in February 1992 on its way to a polar orbit around the Sun.

1997 Cassini-Huygens NASA-ESA flew past Jupiter on its way to Saturn in December, 2000 while Galileo was still operating in Jupiter orbit. That gave a chance to measure the extent of the planets magnetic field, which turned out to be twice the size it had been during the Voyager 1 encounter. Cassini also was able to provide large scale time-lapse images that Galileo could not.

2006 New Horizons NASA Flew past Jupiter on its way to the Kuiper Belt on February 28, 2007. Passing about 2 million miles out, it focused on improving our understanding of the moon system. and is now on its way to Pluto.

012.2 Saturn

Surface

Saturn is somewhat smaller than Jupiter. At 9.58 au from the Sun, it receives little heat from the Sun and has a surface temperature of 270 degrees below zero Fahrenheit. Its surface is heated almost entirely from within, by the gravitational compression heat left over from the original formation of the planet. It is made of much the same material as Jupiter, mostly hydrogen and helium. There are bands and vortexes similar to Jupiter, but they are concealed by a high ice-cloud layer.

One feature unique to Saturn is a hexagonal cloud system at its North Pole.

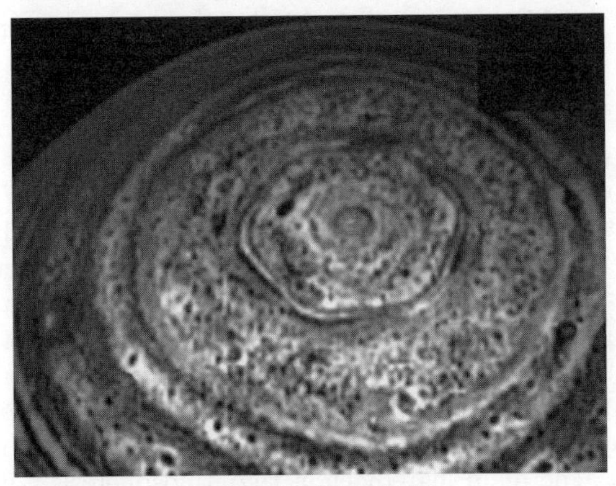

Image: NASA/ESA

The feature was first seen by the Voyager probes in 1980 and is shown here in a picture taken by Cassini in 2006.

Orbit and Rotation

Saturn rotates on its axis once every 10.5 hours. It is difficult to be sure of the rotation rate of a planet with no solid surface and there are still uncertainties. The Voyager space probe tracked radio signals from the upper atmosphere and got a rotation period of 10 hours and 39 minutes. The Cassini probe used a magnetometer to track variation in the magnetic field, which originates deep inside the planet, and got a rotation period that is 8 minutes longer. Like Jupiter, the rotation rate is so rapid that the planet bulges noticeably at the equator and is flattened at the poles. Its rotation axis is tilted at 27 degrees from the perpendicular so there are seasons similar to those on Earth.

One observable effect of the tilt is that we see the ring system at different angles as Saturn goes through its orbit. At the Saturnian spring and fall equinoxes, we see the rings edge-on and they seem to vanish.

Its orbit is elliptical with a perihelion distance of 9.01 au and an aphelion distance of 10.06 au. and takes it around the Sun every 10,759 Earth days (29.46 years).

Magnetic Field

Saturn has a magnetic field which is much weaker than Jupiter's and has about the same intensity as the Earth's field. It is thought that the field is generated by the same type of liquid metallic hydrogen core that Jupiter appears to have. That is consistent with models of its interior which show that its smaller size predicts a much smaller liquid metallic hydrogen core than Jupiter has.

Unlike the magnetic fields of the Earth and Jupiter, the magnetic North and South poles are almost exactly aligned with the planets rotation axis. Without magnetic poles that are being carried around with the planets rotation, it is difficult to get a reliable measure of the planets actual rotation rate.

0.0.3 Moons and Rings

Unlike Jupiter, which has four rather large moons, Saturn has just one large moon, Titan, and a collection of smaller ones. We will mostly be looking at Titan. It is the second largest moon in the solar system and is larger than the planet Mercury. It has a dense and complex atmosphere that obscures its surface when viewed with ordinary light. The atmospheric pressure on the surface is about 60% higher than on Earth. That pressure, combined with a temperature of 289 Fahrenheit degrees below zero means that the atmosphere is related to the triple point of methane in the same way that Earth's atmosphere is related to the triple point of water, so we can expect methane rain, methane snow and methane oceans. The main ingredient in the atmosphere is Nitrogen, just as on Earth. It is possible that the planet harbors some form of life and even if it does not, it offers a deep-frozen version of the early atmosphere of Earth.

Saturn has an extensive system of bright rings that consist mainly of ice-covered rocks. Galileo was the first to notice them, although his small telescope made it look as if the planet had 'handles' on it. Here, the Cassini space probe shows the rings looking back toward Earth.

Image: NASA/JPL/ESA

From the distant vantage point of Saturn, Earth is very close to the Sun and would not ordinarily be visible. Here Cassini took advantage of Saturn blocking the Sun.

From Earth, the rings look like solid bands with a few gaps, the largest one being the Cassini Division. Here, the Cassini spacecraft looks at the gap that it was named after:

Image: NASA/JPL

It can be seen that the rings actually consist of thousands of ringlets with some even occurring within the gaps.

Space Probes

1973 Pioneer 11 NASA Used a gravitational slingshot maneuver at Jupiter to reach Saturn and passed within 21,000 kilometers of its surface on September 1, 1979. It was sent on a dive through the ring plane to see if it was safe for the later Voyager mission to do that. It nearly collided with one of the 'square dancing moons' epimetheus and Janus, coming within a few thousand kilometers of it. Those two moons share the same orbit and exchange positions whenever one overtakes the other.

1977 Voyager 1, Voyager 2 NASA used the Jupiter gravitational slingshot to arrive at Saturn on November 12, 1980 and August 25, 1981. Voyager 1 was aimed at passing Titan and followed a trajectory through the ring plane and then out of the plane of the solar system. Voyager 2 remained in the plane of the solar system on a gravitational slingshot path that would take it on to Uranus and Neptune.

1997 Cassini-Huygens NASA-ESA used the Jupiter gravitational slingshot to enter Saturn orbit on July 1, 2004. The European Space Agency's

Huygens probe dove into Titan's atmosphere and landed on the surface on January 2005, giving the first real look beneath the clouds. It took pictures during its descent by parachute, showing what appear to be drainage channels and a shoreline. the Cassini orbiter continues to explore Saturn's Moon system.

Future missions are still being discussed. The next likely one would launch in 2015 and send probes down into the atmosphere of Saturn.

012.3 Uranus

Surface

Uranus from Voyager 2. NASA/JPL

Uranus is small as gas-giant planets go and is only about four times the size of the Earth. At 19.2 au from the Sun, it receives little heat from the Sun and has a surface temperature of 357 degrees below zero Fahrenheit. It is made of much the same material as Jupiter and Saturn, mostly hydrogen and helium with some methane. The methane provides the blue color of the planet by filtering red light out of the light reflected from a layer of clouds near the top of its atmosphere.

Orbit and Rotation

Uranus rotates on its axis once every 17.24 hours. The rotation axis lies almost in the plane of the ecliptic with Uranus's north pole dipping 7.86 degrees below the plane. Thus, Uranus is actually rotating *backwards* relative to all of the other planets. That tilt makes for extreme seasons, particularly when one of the poles is pointing toward the Sun and the other is pointing away.

Its orbit is elliptical with a perihelion distance of 18.29 au and an aphelion distance of 20.096 au. and takes it around the Sun every 30,687 Earth days (84.02 years). Because we have to wait for 84 years to see a full cycle of seasons, we are not sure just how they manifest themselves. For example, the Hubble Space Telescope image shown later on revealed a sudden abundance of cloud features in 2005 that had not been seen before and are thought to be related to seasonal storms.

Magnetic Field

Uranus has a sizeable magnetic field, which, in keeping with its odd sideways rotation axis, has the magnetic north and south poles tilted at 60 degrees to the rotation axis and displaced from the center of the planet by about a third of its radius. The origin of the magnetic field remains a bit of a mystery because the planet is not large enough to have a liquid metallic hydrogen core like Jupiter and Saturn and there is certainly no liquid iron there. It has been proposed that the interior is basically made of ammonia and water ices, but the structure is not settled.

0.0.4 Moons and Rings

Uranus has 27 satellites that are known so far, with five major ones: Miranda, Ariel, Umbriel, Titania, and Oberon. These are relatively small moons. Titania, the largest one, is only half the size of our own Moon.

Uranus has a ring system that consists of dark rocks. The rings can be seen, along with some surface clouds in this Hubble Space Telescope picture.

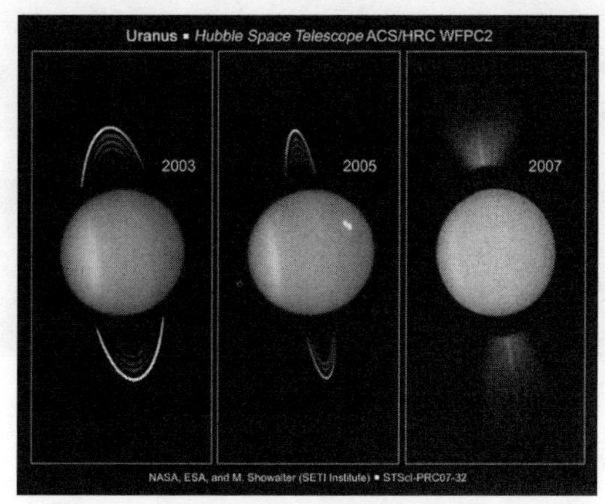

NASA/STSci

Space Probes

Uranus has been visited by only one space probe, **Voyager 2,** which arrived at Uranus after gravitational slingshot maneuvers past Jupiter and Saturn. The closest approach to Uranus was 81,500 kilometers from the planet on January 24, 1986. Voyager 2 discovered 10 additional Moons of Uranus, explored its ring system (which had previously been detected from Earth) and found that the Moon Miranda is extremely odd.

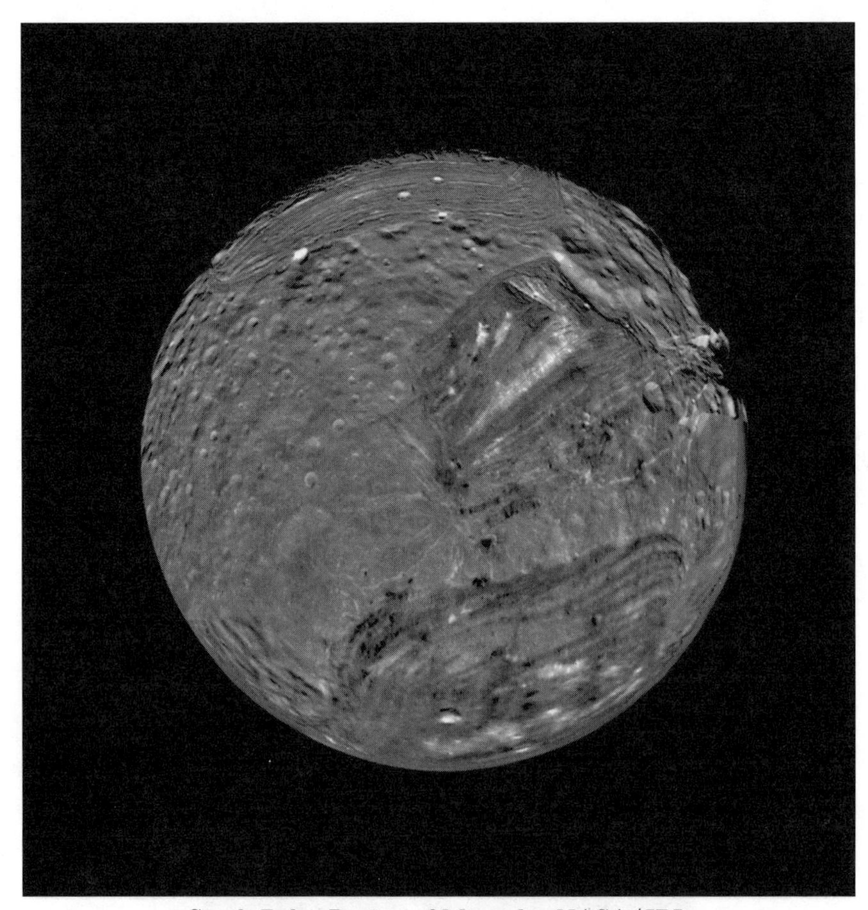

South Polar Region of Miranda: NASA/JPL

The terrain, including the peculiar 'chevron' feature is not like anything else in the Solar System.

012.4 Neptune

Surface

Neptune is approximately the same size as Uranus but considerably more massive.

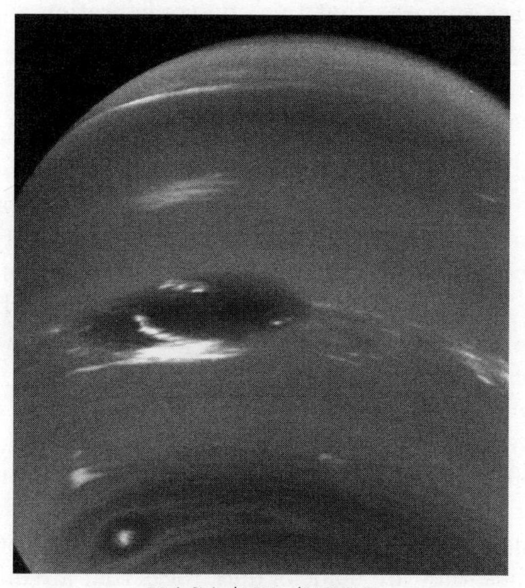

NASA/JPL/USGS

Like Uranus, methane in the atmosphere accounts for the blue appearance. At 30.1 au from the Sun, the intensity of sunlight is very low and the surface is one of the coldest in the Solar System, 353 degrees below zero Fahrenheit. It shows surface features that are somewhat like those of Jupiter, with a Great Dark Spot showing during the 1989 Voyager 2 flyby along with visible bands and clouds. Later, in 1994, the Hubble Space Telescope did not see the spot but found a similar one in the other hemisphere.

The weather on Neptune is extremely violent, probably driven by heat from its interior. Winds reach speeds of 2100 kilometers per hour.

Also like Uranus, Neptune is probably made mostly of frozen gas and water ice. It has been proposed that these two planets are better described as *ice-giants* rather than gas-giants.

Orbit and Rotation

Neptune rotates on its axis once every 16.11 hours. The rotation axis has a tilt of 29.58 degrees from the perpendicular to its orbit, making it similar to Saturn and the Earth in that respect.

Its orbit is elliptical with a perihelion distance of 29.81 au and an aphelion distance of 30.33 au. and takes it around the Sun every 60,190 Earth days (164.79 years).

Magnetic Field

Neptune has a sizeable magnetic field, which has the magnetic north and south poles tilted at 47 degrees to the rotation axis and displaced from the center of the planet by about a .55 of its radius. Thus, it is quite similar to the magnetic field of Uranus, which suggests that this odd configuration is not connected to the extreme tilt of Uranus's rotation axis.

0.0.5 Moons and Rings

Neptune has 13 satellites that are known so far, with just one, Triton, that is large enough to be spherical. Triton is about 80% the size of our Moon with about 30% of its mass, which suggests that it is made mostly of ice. Triton is in a very close *retrograde* orbit around Neptune with a period of just 5.877 days. It is the only major moon in the solar system that is orbiting backwards, indicating that it is actually a captured dwarf planet from the Kuiper Belt that lies just beyond Neptune.

The Moon Nereid was the second moon of Neptune to be discovered and is notable for its extreme orbit. It has a periapsis of 1,372,000 kilometers and an apapsis of 9,655,000 kilometers. That extreme elliptical orbit may be the result of a past encounter between Neptune and another object, possibly Triton.

Like the other Jovian planets, Neptune has a system of rings. Like the rings of Jupiter and Uranus, they are made of dark rocks that are difficult to see from Earth.

Space Probes

Neptune has been visited by only one space probe, **Voyager 2,** which arrived at Neptune after gravitational slingshot maneuvers past Jupiter, Saturn, and Uranus. Because this was the last planet on the tour, the trajectory could be chosen to approach the planet as closely as possible without concern about where the spacecraft would go after that. The closest approach to Neptune was just 3000 miles from the North pole of the planet on August 25, 1989. Voyager 2 obtained some particularly good images of Neptune's ring system.

The next visit to Neptune and its moon Triton is a Neptune orbiter and Triton lander tentatively proposed for 2030.

012 Spot Check

Here are some questions to check your understanding of the material in module 012. Both the answers and where to find these questions at the website may found at the end of the Study Guide.

1 Which of the following objects has a magnetic field with about ten times the intensity of the Earth's field?

a. Saturn

b. Earth's Moon

c. Jupiter

d. Mars

e. Mercury

2 Uranus has

a. an atmosphere of Hydrogen and Helium with some methane.

b. an atmosphere of Hydrogen and Helium with no real surface.

c. an atmosphere of carbon dioxide with about 90 times the surface pressure of Earth's.

d. almost no atmosphere.

e. an atmosphere of carbon dioxide with about 1% the surface pressure of Earth's.

3 Neptune has

a. an atmosphere of carbon dioxide with about 90 times the surface pressure of Earth's.

b. almost no atmosphere.

c. an atmosphere of Hydrogen and Helium with some methane.

d. an atmosphere of carbon dioxide with about 1% the surface pressure of Earth's.

e. an atmosphere of Hydrogen and Helium with no real surface.

4 Which of the following objects is the largest moon of Saturn?

a. Titania

b. Ganymede

c. Titan

d. Callisto

e. Triton

5 Saturn has

a. an atmosphere of Hydrogen and Helium with some methane.

b. an atmosphere of carbon dioxide with about 1% the surface pressure of Earth's.

c. an atmosphere of carbon dioxide with about 90 times the surface pressure of Earth's.

d. almost no atmosphere.

e. an atmosphere of Hydrogen and Helium with no real surface.

6 The first spacecraft to go into orbit around Saturn was

a. Galileo

b. Pioneer 11

c. Mariner 9

d. Cassini-Huygens

e. Ulysses

7 Voyager 2 is the only space probe so far to have visited

a. Mercury

b. Venus

c. Jupiter

d. Saturn

e. Uranus

8 Uranus's magnetic North and South poles are

a. displaced from its rotation axis poles and also from the center of the planet.

b. displaced from its rotation axis poles, but still symmetrical about its center, much like Earth's.

c. almost exactly aligned with its rotation axis.

d. near the equator of the planet.

9 Which of these planets has a rotation axis that is inclined in much the same way as that of Earth?

a. Mercury

b. Saturn

c. Jupiter

d. Uranus

10 Which of the following planets has at least 62 moons?

a. Mars

b. Venus

c. Earth

d. Neptune

e. Jupiter

11 Which of the following spacecraft is the only one to have flown past Neptune?

a. Galileo

b. Voyager 1

c. Pioneer 11

d. Cassini-Huygens

e. Voyager 2

12 Pioneer 10 was an early space probe sent to fly past

a. Mercury

b. Earth's Moon

c. Jupiter

d. Mars

e. Venus

13 Which of the following planets is 11 times the size of the Earth?

a. Jupiter

b. Mars

c. Venus

d. Uranus

e. Neptune

14 Uranus rotates on its axis once in about

a. 10 hours.

b. 10 Earth days.

c. 88 Earth days.

d. 17 hours.

e. 24 hours.

15 Which of the following objects has a magnetic field whose North and South poles are displaced from its rotation axis poles and also from the center of the planet?

a. Saturn

b. Jupiter

c. Neptune

d. Venus

e. Mercury

013: Comets and the Outer Solar System

013.1 Comets in Detail

Comet Tails

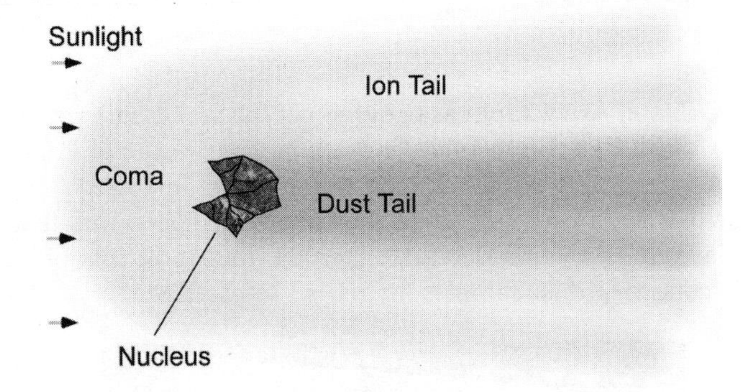

A comet is a ball of frozen gas and ice with some dust mixed in. Comets have been referred to as 'dirty snowballs.' They originate far from the Sun and follow highly elliptical orbits.

The characteristic tail of a comet is formed by the action of sunlight.

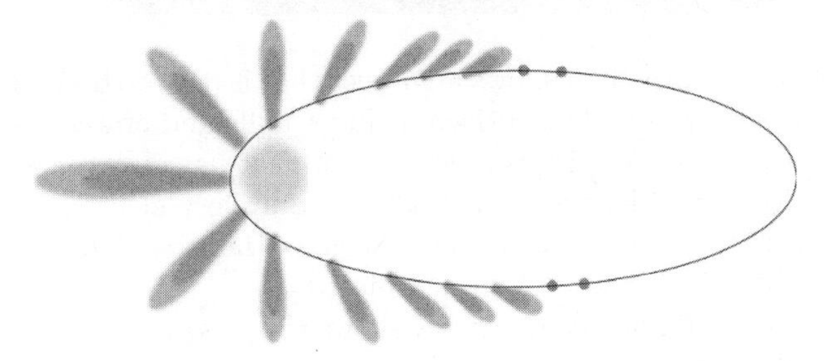

While they are farther from the Sun than the planet Mars, comets are simply balls of frozen gas and dust and do not have tails. Closer to the Sun, the gas evaporates and the dust that was held together by the frozen gas is released. The Solar wind pushes these materials away from the Sun to form tails. Notice that the tail leads the comet on its trip away from the Sun.

The *Ion Tail* consists of atoms that have gained or lost electrons and consists of straight streamers. It forms first and can be as long as the size of the Earth's orbit.

The *Dust Tail* is usually fuzzy looking and is curved because the individual dust particles are taking up their own separate orbits around the Sun. It forms closer to the Sun.

Parts of a Comet

Nucleus	The 'dirty snowball'	a few kilometers across
Coma	Vaporizing gas and dust	as large as the planet Jupiter
Ion Tail	Atoms with electrons missing	as large as the Earth's orbit
Dust Tail	rock fragments	Forms close to the Sun

Examples

At the website (http://www2.jpl.nasa.gov/comet/pach17.html) you will find an image of the Comet Hale Bopp taken in Joshua Tree National Park by Wally Pacholka on April 7, 1997. Notice the ion tail and the dust tail.

As we saw earlier, in module 10.5, The Stardust spacecraft flew into the coma of a relatively new comet, Wild-2 in 2004, and took this picture of the nucleus while collecting dust samples for return to earth:

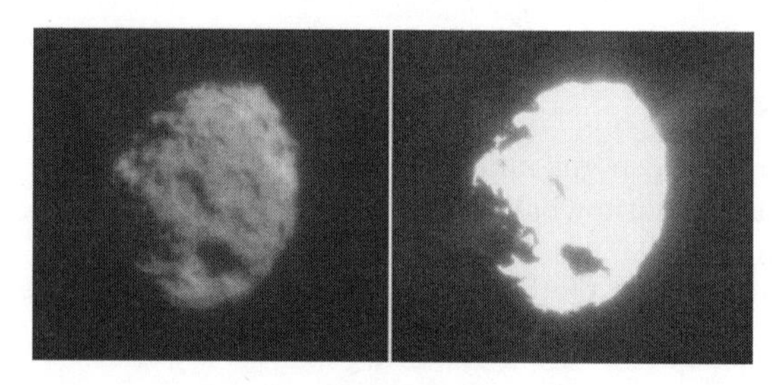

The short exposure, on the left, shows an object that is not too different from the moons of the outer planets. The surface includes cliffs and craters, indicating a fairly strong outer crust. Wild-2 is a new comet that has only made about five Sun passes so far and it does not approach the Sun very closely, so its surface has not yet been modified. Older comets, such as Halley's Comet, have nuclei that look very different, as we saw in module 10.5.

The longer exposure, on the the right, shows the jets of gas that are forming the coma of the comet.

013.2 Meteor Showers

Language Lesson

Before it hits the atmosphere it is a	*Meteoroid*
Passing through the atmosphere it is a	*Meteor*
Once it has landed, it is a	*Meteorite*

Meteor Showers occur when the Earth passes through the orbit of a shattered comet. The point in the sky where the shower seems to come from is called the *radiant* and indicates the direction of the Earth's motion when it encounters the cloud.

Because the Earth is always moving in the same direction in space when it passes through a given comet orbit, the radiant is always the same. Thus, meteor showers are named after the constellation that contains their radiant.

Because a given comet orbit is always encountered at the same point along the Earth's orbit, each meteor shower always happens at the same time of the year.

Examples

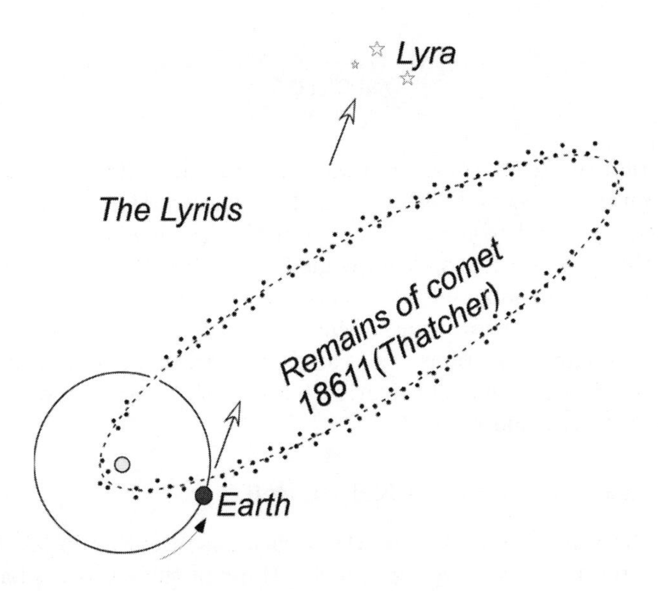

The *Lyrid shower* seems to come out of the constellation Lyra and peaks in April. This year it is the night of April 21–22, 2008 and will not be very visible because it coincides with a full moon. The comet that gave rise to this shower was comet 18611 (Thatcher).

The best known meteor shower is probably the *Leonid shower*, which peaks in November and has its radiant in the constellation Leo. This year its peak is the morning of November 17, 2008.

013.3 The Origin of Comets

Long Period Comets and the Oort Cloud

Some comets arrive near the sun on trajectories that will not bring them back again for millions of years. These *long period comets* come from random directions, not just in the plane of the solar system. A somewhat arbitrary definition of a long-period comet is that its orbital period is greater than 200 years.

In 1950, Jan Hendrick Oort was trying to explain how a solar system that is billions of years old can still have comets when each comet is destroyed after only a few orbits past the Sun. His answer was that there is a very large reservoir of potential comet nuclei in a cloud at the far fringe of the Solar System. The cloud supplies new comets as fast as the old ones are destroyed.

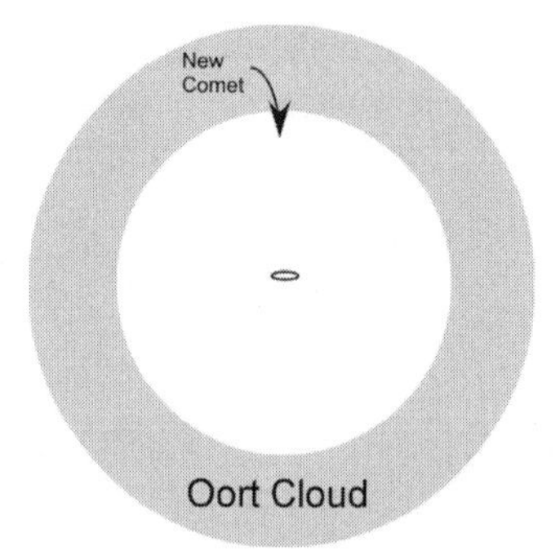

The question that Oort considered was not new and neither was his answer. Both had been discussed by Ernst Öpik in 1932 and probably by others before that. However, we tend to name things after the last person to present them rather than the first, so the cloud has come to be called the *Oort Cloud.*

When the paths of long-period comets are traced backward in time, it is found that they have fallen in from about 50,000 astronomical units out. Since they come in random directions, the cloud appears to be spherical. At the present time it is somewhat speculative because there are almost no direct observations of Oort cloud objects.

Short Period Comets and the Kuiper Belt

Comets that have periods that bring them back near the Sun in less than 200 years are referred to as *short period* comets. Many of these objects have orbits that are approximately in the same plane as the orbits of the major planets. The Kuiper Belt is a reasonable candidate for the origin of many of these comets. Unlike the Oort cloud, we have begun to observe large numbers of Kuiper Belt objects, so it has become a well-tested concept.

Unlike the Oort Cloud, the Kuiper Belt is of limited extent. The limits are determined by orbital resonances with Neptune. At the inner edge of the main belt are objects whose orbital periods are in a 3:2 relation to the orbital period of Neptune. The outer edge is defined by objects whose orbital periods are just twice the orbital period of Neptune. This *Main Kuiper Belt* goes from 39.5au at its inner edge to about 48au at its outer edge. The *fully extended Kuiper Belt* extends a bit farther, from 30au to 55au.

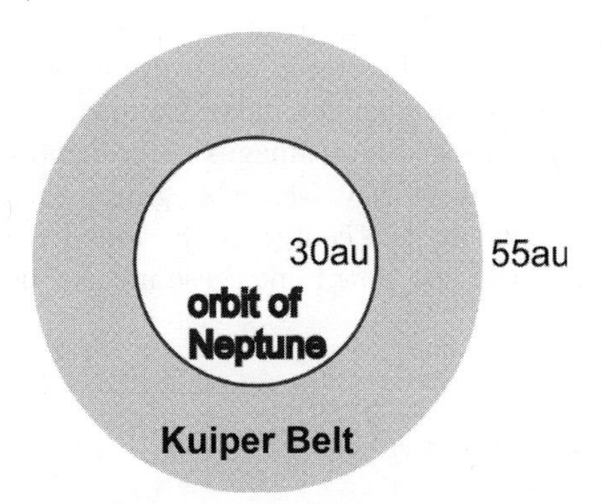

Yet another variation on the Kuiper Belt is the region that is largely uninfluenced by the gravity of Neptune. That region extends from 42 to 48au and is called the *Classical Kuiper Belt.* About 2/3 of the thousand or so Kuiper Belt objects found so far are in that region.

Trans Neptunian Objects

As the number of objects discovered has increased, new classes of objects have been defined and named. Here is a partial list:

TNOs or Trans-Neptunian Objects — Objects beyond the orbit of Neptune.

KBOs or Kuiper Belt Objects — objects in the extended Kuiper Belt

SDOs or Scattered Disc Objects — objects beyond the Kuiper Belt out to about 100au.

Plutinos — About 200 objects in the same type of orbit as Pluto. These objects orbit the Sun twice for every three times that Neptune does. They are said to be in 3:2 resonance with Neptune.

Cubewanos — Objects in the Classical Kuiper Belt. Named after the first such object, 1992QB1.

Planemos — Planetary Mass Objects. These are large enough to be spherical but are not necessarily in orbit around the Sun or any other star. They include possible 'rogue planets' that drift between the stars entirely on their own.

Examples

Most of these objects are in the main Kuiper Belt and the Scattered Disk. The one exception is *Sedna*, which has a highly elliptical orbit with its perihelion at 76au and its aphelion at 975au. It was discovered in 2003 while it was approaching perihelion. It is regarded as an inner Oort cloud object.

All of these objects were found in essentially the same way that Pluto was found by Clyde Tombaugh in 1930. The method is to take pictures of the same part of the sky at different times and then switch back and forth between the two pictures with all of the star images lined up so that they do not change. Any object that is not a star will move from one picture to the next and will be seen to be jumping back and forth as the images are alternated.

Tombaugh used a mechanical device called a 'blink comparator' with a flipping mirror to alternate images. The rapid discovery of other Trans Neptunian objects came only after telescopes using Charge Coupled Devices (CCDs) came into common use. The resulting digital images made it possible for computers to search for the few faint images that were changing position. The discovery picture of Sedna with Sedna at the center of the green ring in the animated image on our web site, shows just how faint these images are.

013.4 The Transition from Kuiper Belt to Oort Cloud

Our current picture of the region beyond the Kuiper Belt is shown in this picture:

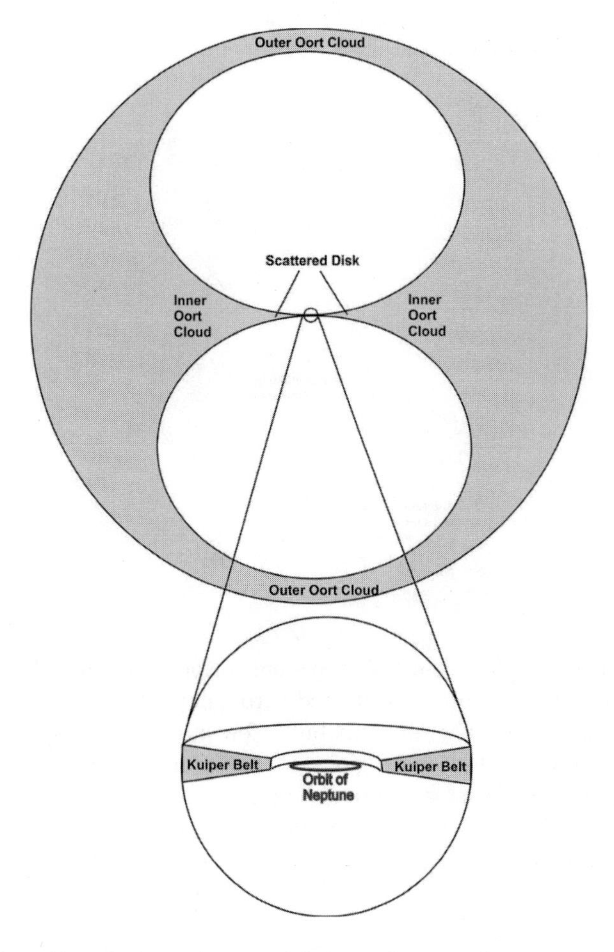

The *Inner Oort Cloud* is a wide toroidal shaped region from 50 to 20,000au that starts where the Kuiper Belt stops. The innermost part of this region, out to about 100au is often referred to as the *Scattered Disk*. Sedna, with its highly elliptical orbit that reaches almost 1000au is regarded as a visitor from the Inner Oort Cloud.

013.5 Beyond the Oort Cloud

The Oort cloud reaches out about one light-year, about a quarter of the way to the nearest star. The assumption until recently has been that the space between the Oort clouds of neighboring stars is mostly empty. However, that might not be true.

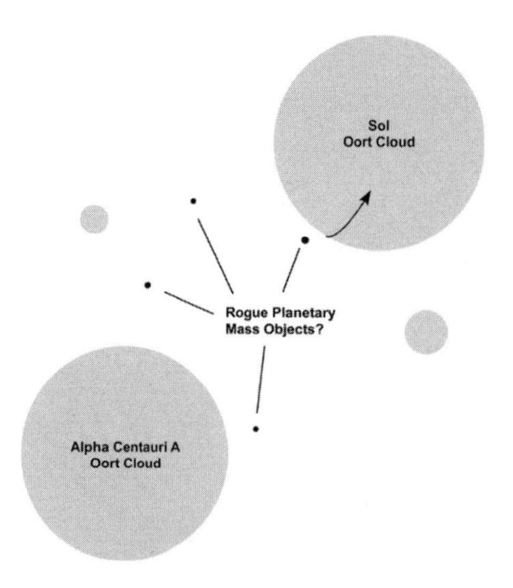

Our current theory of how the Solar System formed suggests that there may be many planetary mass objects scattered throughout interstellar space. Direct observation of these objects is difficult but a few large, relatively hot ones have been spotted in the star-forming regions of the Orion Nebula.

If rogue planets do exist and are as common as we now suspect, many of them are likely to be similar to the ice planets of the Kuiper Belt and would make ideal base camps for interstellar expeditions at some far future time.

013 Spot Check

Here are some questions to check your understanding of the material in module 013. Both the answers and where to find these questions at the website may found at the end of the Study Guide.

1 The Oort cloud of our Sun reaches

a. beyond several of the nearest stars to our Sun.

b. a negligible part of the distance from our Sun to the nearest star.

c. at least a quarter of the way to the nearest star.

d. most of the way to the nearest star.

2 You hear a loud noise outside and go outside to find a smoking rock embedded in your driveway underneath your wrecked car. The object is probably a

a. meteoroid.

b. comet.

c. asteroid.

d. meteorite.

e. meteor.

3 The Inner Oort Cloud is located

a. in the same general area as Pluto.

b. between the outer edge of the Kuiper Belt and the orbit of Pluto.

c. between the orbit of Neptune and the inner edge of the Kuiper Belt.

d. beyond the Kuiper Belt.

4 The tail of a comet always points

a. in its direction of motion.

b. away from the Sun.

c. toward the Sun.

d. toward the Earth.

e. opposite to its direction of motion.

5 The objects of the Kuiper belt are mostly orbiting

a. beyond all of the Jovian planets.

b. among the Jovian planets.

c. between the orbits of Earth and Mars.

d. within the asteroid belt.

014: Formation of the Solar System

014.1 The Solar Nebula

The Sun is a Typical Star

We will discuss stars in detail later. Our Sun is a typical middle-aged, medium temperature star. Its system of planets also appears to be fairly typical although we have only been able to study a few hundred planets of nearby stars.

Astronomers cannot test their ideas about the formation of the Solar System by trying to create stars and planetary systems in the laboratory. However Nature has provided something that is almost as good: thousands of examples of stars and planetary systems that are forming now.

Interstellar Clouds and Star Clusters

Our solar system began as a slowly rotating cloud of gas and dust. Every part of the cloud is attracted to every other part by gravity, so the cloud becomes smaller until it looks like this example:

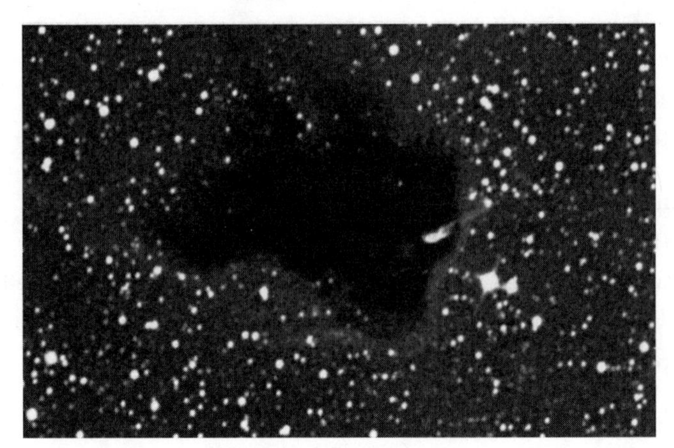

Digitized Sky Survey

Here the cloud is dense enough to hide the stars behind it. At this stage, the cloud is usually referred to as a *globule.* The particular globule that led to the formation of our own Sun and Solar System is called the *Solar Nebula.*

Globules often come in bunches as in Thackeray's Globules shown here:

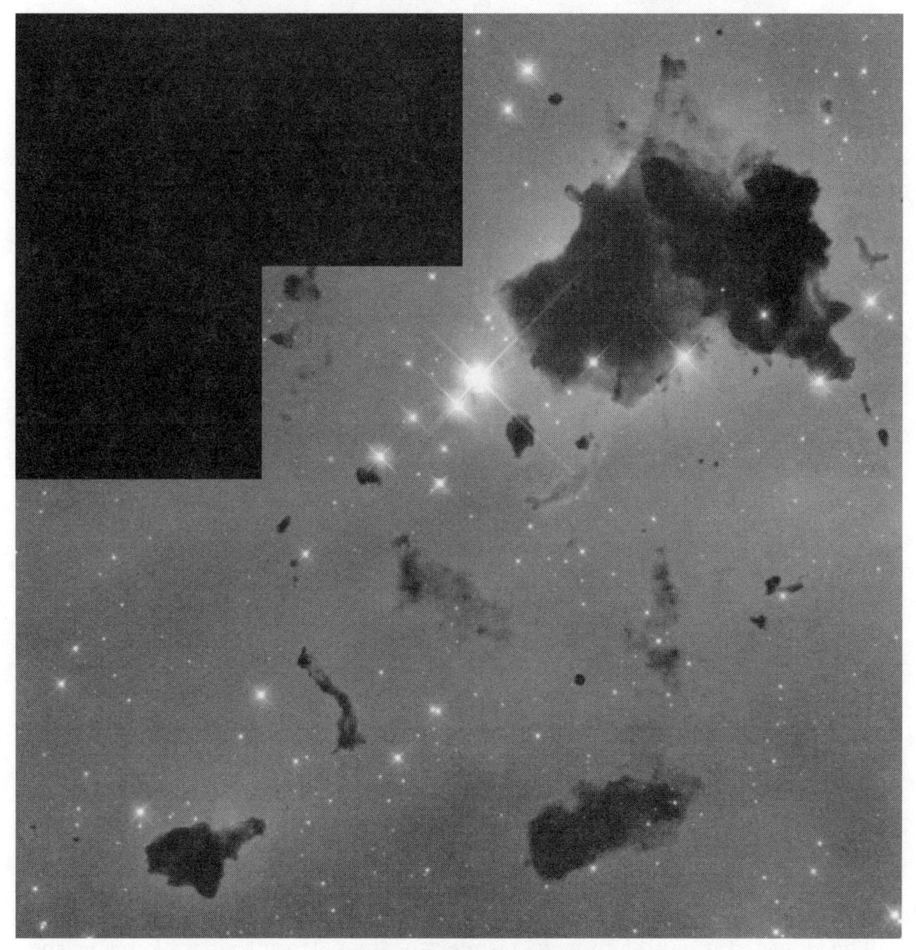

NASA/STSci/AURA

These collections of globules come from the fragmentation of larger clouds and eventually lead to clusters of stars. An extreme example of a large cloud fragmenting and forming many stars is NGC604.

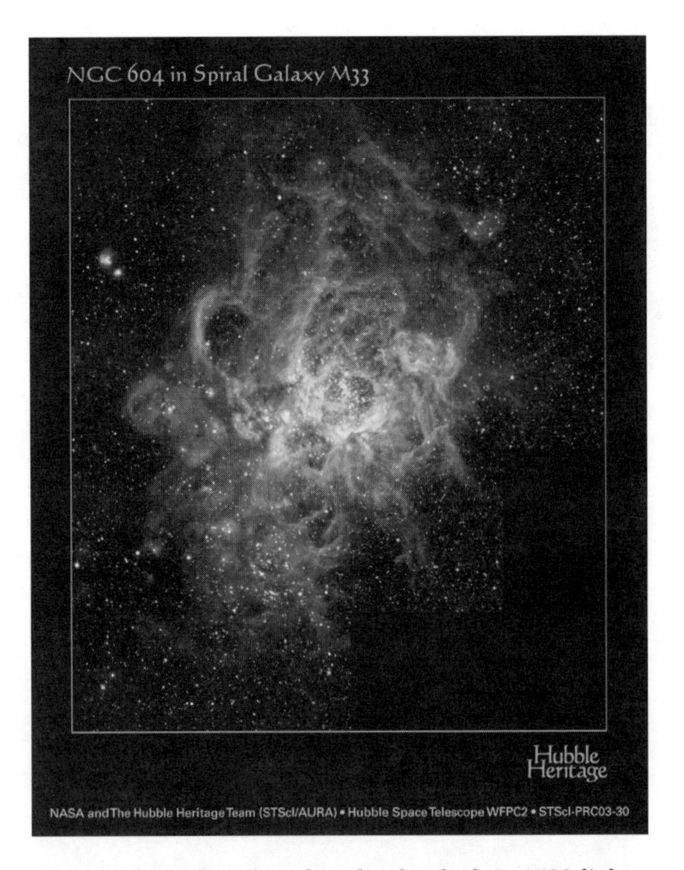

Over 200 stars have formed within this cloud, which is 1500 light years across.

These clusters of stars remain long after the original cloud has vanished. A familiar cluster that still shows wisps of the original cloud is the Pleiades often called the seven sisters. Actually they really are sister stars but there are about 400 of them.

The closest star cluster to us forms most of the pattern that we know as the Big Dipper. All but the end stars of the Dipper belong to a single cluster that is moving through space together and passing very near our Sun.

Rotation

The orginal Solar Nebula, was rotating very slowly. As it collapsed, it rotated faster.

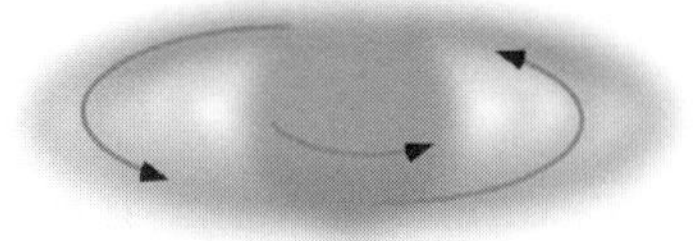

The material that formed the Sun collapsed at the center. In order to collapse, it had to transfer some of its rotation to the rest of the cloud, which condensed into a rotating disk. The planets condensed out of the disk.

As a result, the planets rotate around the Sun in the same direction that the Sun rotates and in the equatorial plane of the Sun.

Because they condensed out of the rotating disk, most of the planets rotate on their axes in this same sense — counterclockwise when viewed from above the Earth's North Pole.

014.2 The Protostar Stage

When the collapsing Solar Nebula reached a density that light could not penetrate, the heat from compressing the cloud became trapped and the central temperature began to rise rapidly.

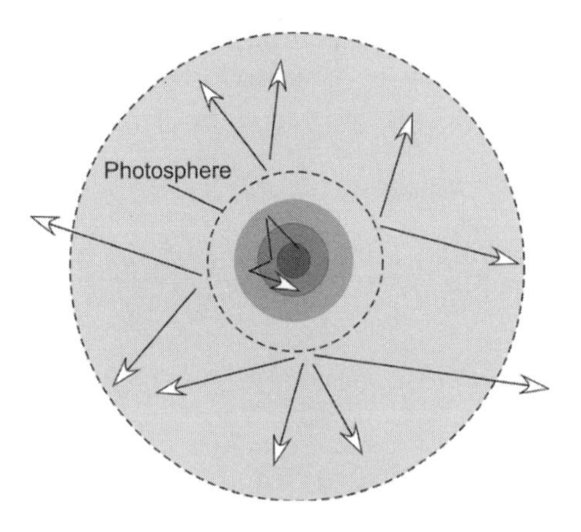

Where the material thins out enough for light to escape, we see the apparent surface or *photosphere* of the new *protostar* or, in this case, the *proto-Sun*.

Examples of Protostars Forming Now

Although protostar surface temperatures are low, their surface areas are very large, so they can be extremely bright. Shining entirely from the gravitational energy released by its collapse, a protostar can be a thousand times brighter than our Sun. Most of that energy comes out at light wavelengths too long for our eyes to see, so we have to look for them with infrared telescopes.

On our website, you will find several pictures of protostars that are forming now. The pictures depend on color, so it is best to look at them on the website.

There is a collection of protostars (the reddish dots) in the R Corona Australis star-forming region, 500 light-years from our Sun. The picture on our website is actually made in infra-red light by the University of Hawaii 88 inch telescope. The colors are "false" because our eyes are not sensitive to any of the wavelengths being detected by the telescope.

There are protostars forming 1000 light years away in the Serpens Region, again in an infra-red picture shown on our website.

The Elephant's Trunk Nebula in the constellation Cepheus is shown in ordinary light and then in infrared. The nebula looks very different when viewed by the Spitzer Space Telescope in infrared light. The bright reddish objects in the picture are protostars.

014.3 Condensation of the Planets

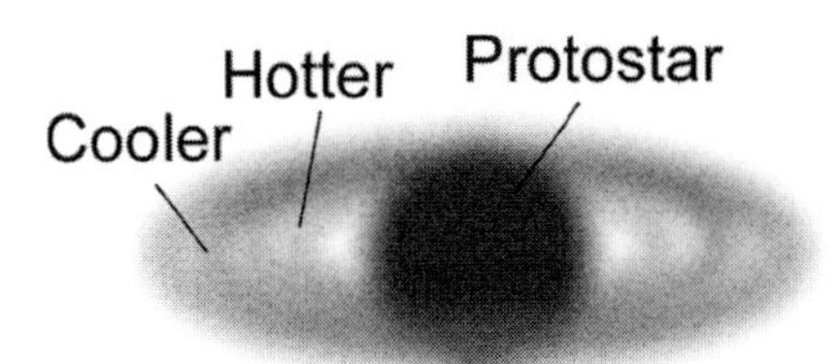

Once the protoSun formed, any ices that had formed earlier vaporized in the hot inner region. Only rock and iron grains survived there and gradually collected into larger objects that eventually became the terrestrial planets. Since all of the water ice and frozen gases had been driven away from this region by the heat, these proto-terrestrial planets would have been completely without water and without atmospheres.

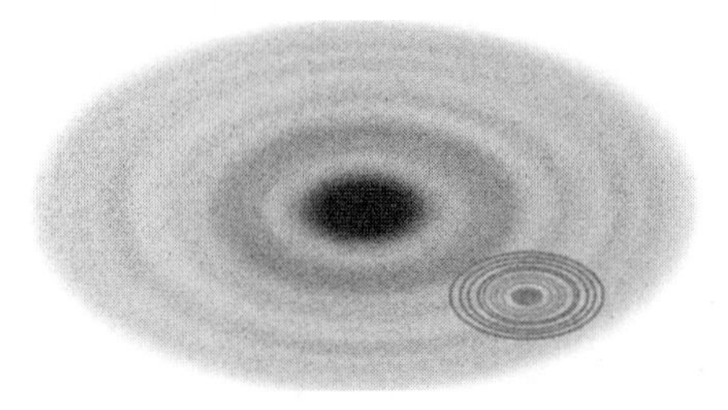

Farther out in the Solar Nebula, the original icy fragments were collapsing into a few miniature Solar Nebulae, each forming into a miniature Solar System with a large central object and smaller objects condensing out of a disk around it. These were the beginnings of the Jovian planets.

014.4 Jupiter Loses the Race

The Tau-Tauri Wind

In the early Solar System, the Sun and the Jovian planets were essentially similar objects, each large enough to grow still larger by collecting gas from the Nebula. Each with a rising central temperature.

The Sun won the race. When it began the final collapse that would raise its central temperature into the range needed to ignite the nuclear fusion reaction that powers stars, its outer layers blew off.

The result is called a *Tau Tauri wind*, named after the first protostar that was seen to be in that stage of its evolution. The Tau-Tauri wind blew all of the remaining gas out of the solar system so that Jupiter and the other Jovian planets could not grow any larger and would never become stars.

0.0.1 Binary Star Systems

Half of the stars in the sky are actually binary or even higher multiple systems. In these cases the 'Jovian planets' managed to grow fast enough to ignite before all of the gas was blown away by a Tau Tauri wind. Why they made it while Jupiter failed is not exactly known.

It was once thought that binary star systems would not be able to form Earth-like planets. Nature has shown us by example that this idea was wrong. Here is what we think is now happening in the newly formed binary star system HD113766, 424 light years away from us.

The brown ring in this artists image represents dusty material detected in this system. It is thought to be enough to form a terrestrial planet at exactly the right distance from its sun to have liquid water. The white ring represents icy material, also detected around this star that could be a possible source of that water. A few billion years from now, that system could host an Earthlike planet. The implication is that many binary star systems could have Earthlike planets.

That is particularly interesting because the closest star system to us is a trinary system (Alpha Centauri) that includes a near-twin of our own Sun.

014.5 The Fates of Dirty Snowballs

Unstable Orbits

Out beyond where the terrestrial planets were forming, a large number of icy objects — dirty snowballs — condensed out among the Jovian planets and in the region beyond Neptune. These objects saw a changing gravitational field because of the gravity of the giant Jovian planets. There are very few stable orbits in such a changing field, so all of these objects were ejected from that part of the solar system, some moving outward and some falling inward.

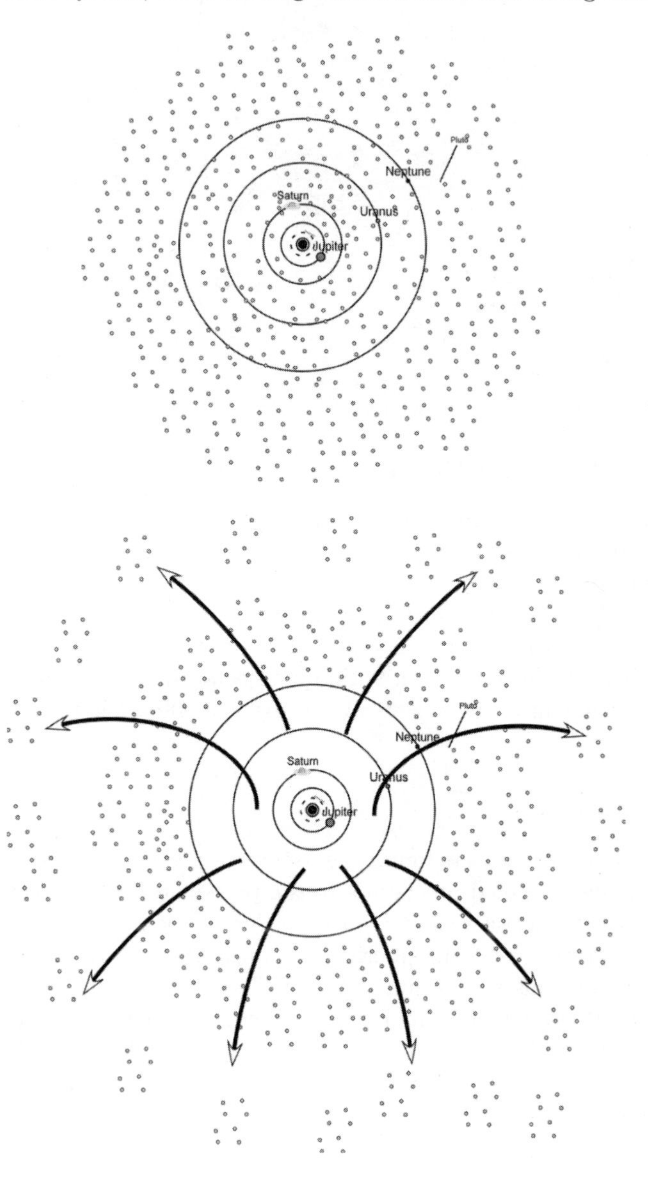

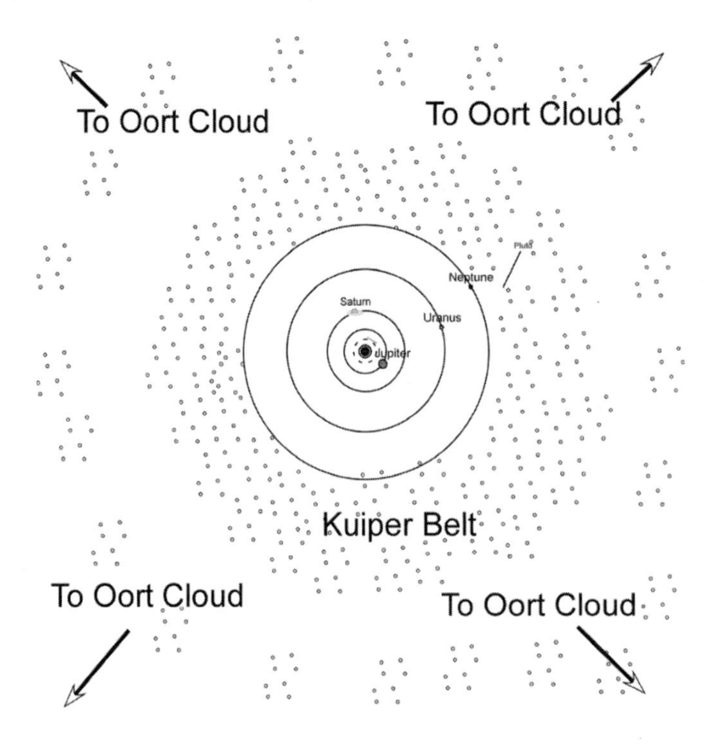

The Origin of the Oort Cloud

The *outward-bound* icy objects were sent in all directions on trajectories that took them far from the Sun, where they formed a spherical cloud of ejected material, the Oort Cloud.

An interesting prediction of this model is that some icy objects would have been sent out at escape velocity from the Solar System and are now wandering the space between the stars along with icy objects from the formation of other stars. That argument is one of several reasons that interstellar space might not be all that empty.

The Rain of Comets

The *inward-bound* icy objects found themselves in a very busy part of the Solar System with newly formed precursors to the present terrestrial planets. These proto-planets were massive enough to sweep up the icy objects. It is thought that all of the water and atmospheric gases on Earth and the other terrestrial planets came from this early "rain of comets."

In addition to water and bits of rock, comets also contain a variety of hydrocarbon compounds that are precursors to organic molecules. Thus, the rain of comets also supplied the building blocks and possibly the initial food supply for living things.

Pluto and the Origin of The Kuiper Belt

The icy objects that formed just *beyond Neptune* had stable orbits and remained there to become the Kuiper Belt. In 1951, Gerard Kuiper noted that such a belt of icy objects would be left in that region. However, at the time that he was working, Pluto was thought to be a full-fledged planet with about the mass of Earth. Such a planet in Pluto's elliptical orbit would have completely disrupted the orbits of any objects in that region, so he predicted that no such objects would be found.

The newly adopted definition of a planet requires that it must have cleared the neighborhood of its orbit. Kuiper was correct in predicting that an Earth-mass Pluto would have cleared its neighborhood of most objects. Actually, Pluto has only 1/500 the mass of the Earth and has obviously not cleared the neighborhood of its orbit, so it is no longer regarded as a planet.

014 Spot Check

Here are some questions to check your understanding of the material in module 004. Both the answers and where to find these questions at the website may found at the end of the Study Guide.

1 The plane that contains the Earth's orbit around the Sun is also called the plane of the ecliptic. When you look for the planets in the sky, you expect to find

a. all of them near the ecliptic.

b. all of them near the ecliptic except for Neptune.

c. all of them near the ecliptic except for Venus.

d. none of them near the ecliptic except for Mars.

e. all of them near the ecliptic except for Uranus.

2 In the original Solar Nebula, objects that condensed near the protoSun tended to be mostly rock and iron rather than volatile gases and water because, in that part of the nebula

a. it was too cold for volatile gases and water to condense.

b. there were no volatile gases and water.

c. it was too hot for volatile gases and water to condense.

d. it was cold enough for rock and iron to condense.

3 Jupiter failed to become a star because

a. The Sun's gravity prevented it from growing.

b. It was too far from the Sun.

c. It was made from the wrong material.

d. The Sun's final collapse blew all the gas away.

e. When Jupiter ignited, the Sun blew it out.

4 Protostars are

a. far less bright than our Sun because of their low temperatures.

b. essentially invisible and almost impossible to detect.

c. about as bright as a typical star such as our Sun.

d. far brighter than typical stars because of their large surface area.

5 The Kuiper Belt is thought to have originated when

a. icy objects condensed out of the interstellar medium.

b. a planet failed to form near Jupiter.

c. icy objects were ejected inward from among the Jovian planets.

d. icy objects condensed out just beyond Neptune.

e. icy objects were ejected outward from among the Jovian planets.

015: Earth's Atmosphere and Interior

015.1 A Thin Layer of Air

The Earth's atmosphere is a thin skin around the planet.

From sea level to the top of the Ozone layer at 65km is only 1% of the Earth's 6500km radius. The total mass of the Earth is 1,200,000 times the mass of its atmosphere. The main components of the atmosphere are

Nitrogen	78%
Oxygen	20%
Water	1%
Argon	0.9%
Carbon Dioxide	0.04%

Notice that Carbon Dioxide is a *tiny* fraction of the atmosphere: one part in 2500.

015.2 Convection

The key to understanding the atmosphere

Warm air rises

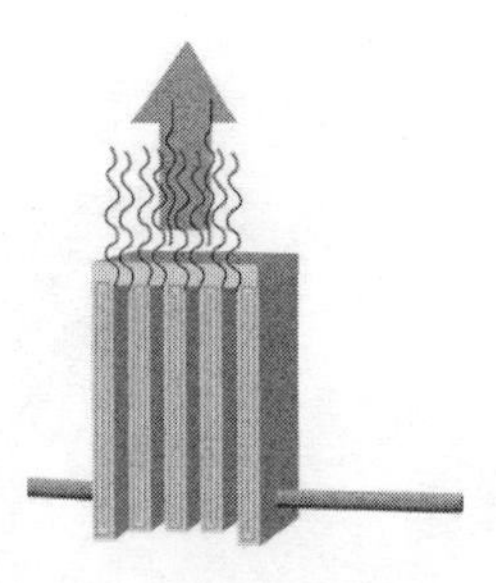

Cold air falls

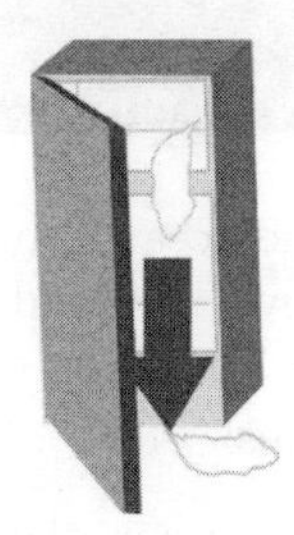

Ocean Breezes

During the day, a wind blows from the ocean toward the land

The heated air over the land rises while the cooler air over the ocean sinks and rushes in to replace it.

Once the sun sets, the wind reverses and blows out to sea.

The land cools off quickly while the ocean stays warm, so the warmer air rises over the ocean and the cooler air over the land sinks and rushes seaward.

Stable and Unstable Air Masses

Where there is cold air on top of warm air, the air on the bottom wants to rise while the air on the top wants to sink. The result is unstable and things start to happen.

Temperature falls as you go up: unstable.

Where there is warm air on top of cold air, the air on the bottom wants to be on the bottom while the air on the top wants to be on the top. The result is stable and just sits there.

Temperature rises as you go up: stable.

015.3 Temperature Layers

Defining Property: Temperature Change With Height

The layers of the atmosphere are distinguished by how the temperature changes with height.

Near the Earth's surface, the Sun warms the air nearest the ground, so the temperature falls with increasing height up to about 12km. This region is called the *troposphere.*

At about 45km the ultraviolet light in the Sun is absorbed in the ozone layer so the temperature rises with increasing height. This region is called the *stratosphere.*

The temperature reaches a maximum and once again falls with increasing height in the *mesosphere.*

Finally, where ultraviolet light knocks electrons off atoms, there is a region where the temperature once again rises with increasing height, the ionosphere.

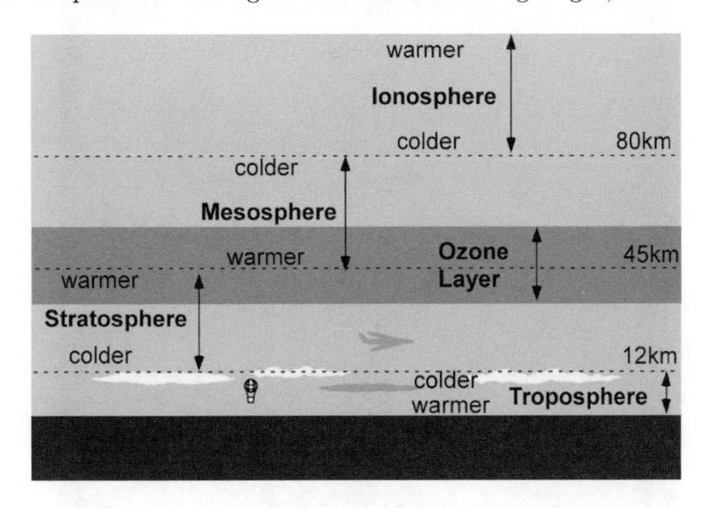

Troposphere

The troposphere is unstable because the warm air near the ground wants to rise. Here is where weather happens.

Stratosphere

The stratosphere is stable because the colder air is at the bottom and the warmer air at the top.

When ash and dust get into the stratosphere, they stay there for years because there is no vertical air circulation there.

The eruption of Mt. Pinatubo created a haze in the stratosphere that reduced the sunlight reaching the Earth's surface by about 5% for several years afterward.

Ozone Layer

The Ozone Layer consists of an unstable form of Oxygen (three atoms instead of the usual two). It shelters the Earth's surface from ultraviolet radiation.

Here are some false-color images showing the large and growing hole in the ozone layer over Antarctica. Conditions over Antarctica exaggerate the effect of chlorofluorocarbons on ozone destruction. For more information on the ozone layer, go to the NASA Earth Observatory.

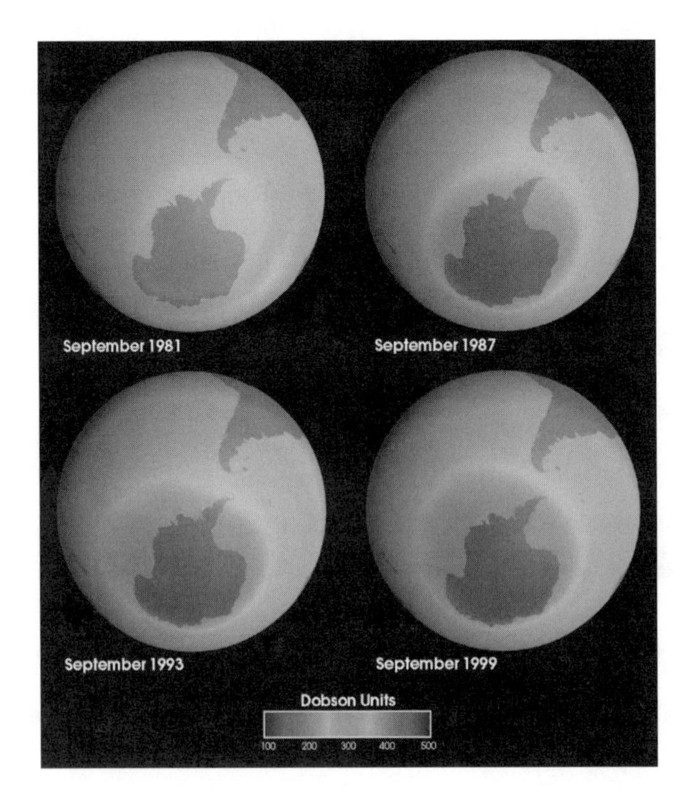

Ionosphere

The Ionosphere, (once called the Heaviside Layer) reflects radio waves and is the basis for long-distance radio transmission.

015.4 Greenhouse Effect

Greenhouse Gases

The Earth's atmosphere is mostly transparent to visible light, which reaches the surface.

The heated surface gives off infrared light, which carries away energy.

The "*Greenhouse gases*," such as carbon dioxide methane, and water vapor absorb infrared light and then re-radiate it in all directions, sending much of it back to the ground.

The more greenhouse gases there are, the higher the Earth's average temperature.

Because Carbon Dioxide is only 1/2500 of the Earth's atmosphere, it is the greenhouse gas that is most sensitive to human activities.

A Short Horror Story

1. We add carbon dioxide to the atmosphere by burning lots of coal (which is almost pure carbon).
2. The greenhouse effect raises the earth's temperature.
3. The polar ice retreats and reflects less light into space.
4. More light absorbed by the earth raises the earth's temperature some more.
5. Higher temperatures mean more water vapor in the air, another greenhouse gas.
6. Go back to step 2.

This scenario is sometimes called the *runaway greenhouse effect.*

Would it happen that way?

The "runaway" scenario assumes that our climate is basically *unstable* and can be disrupted by a small shove in the wrong direction.

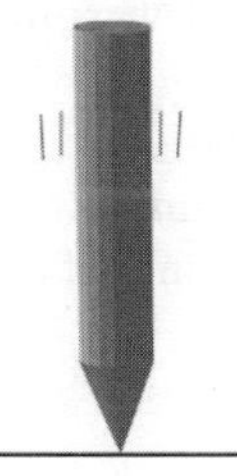

There are many "restoring forces" that would counter an enhanced greenhouse effect.

- The oceans absorb huge amounts of carbon dioxide, so most of the increase might not end up in the air.
- Increased water vapor means more cloud cover and less light being absorbed by the earth.
- Higher temperatures mean longer growing seasons which cause plants to absorb more carbon dioxide.

Would these restoring forces be enough? They might be enough if we do not push the climate too far. That would correspond to our climate being a *metastable* system like a pencil balanced on its flat end.

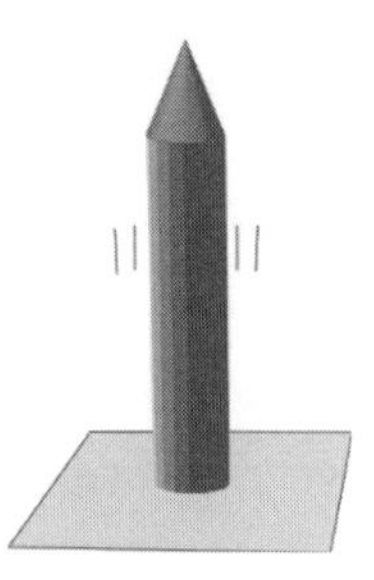

In that case, however, too large a shove will still knock the pencil over, corresponding to a runaway scenario.

Global Warming and Climate Models

Most statements about the effects of human activities on the climate are based on our best models of how the atmosphere works. The Earth's climate is an extremely complicated system, so the models could be leaving out something important.

For example, the earliest models assumed that the atmosphere can be divided into layers at different altitudes with everything in a given layer being the same. That sort of "spherical chicken" model may capture the essential features of the situation, but leaves out important details such as continents and oceans. Gradually more and more detailed climate models have been developed and so far, the essential prediction of global warming from carbon emissions remains the same.

As with anything in science, the validity of the prediction will remain in doubt until we actually do the experiment. We are, of course, doing the experiment now by putting increasing amounts of carbon dioxide into the atmosphere.

015.5 The Earth's Interior

The Layers

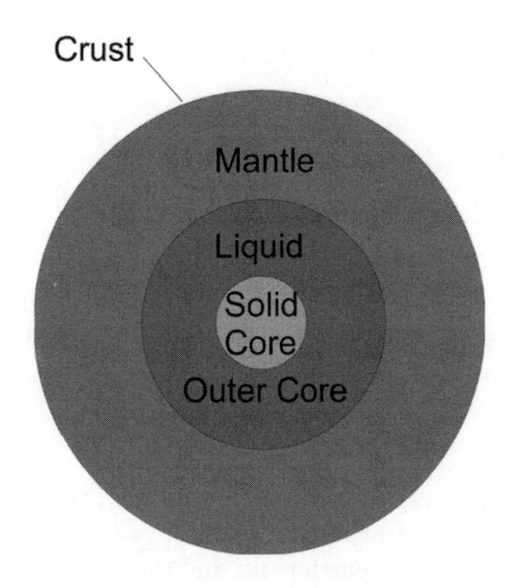

- The *Crust* is made of low density rock such as granite and is only about ten miles thick.
- The *Mantle* is made of denser rock that is semi-liquid in constant slow motion.
- The *Core* appears to made of iron and nickel. The outer core is liquid while the inner core is solid.

The material of the mantle behaves like a stiff solid when pressure is applied suddenly, but flows like a liquid under gradual pressure. The most familiar example of that sort of material is silly putty, which bounces like rubber but will gradually flow through a hole if you give it enough time.

Seismic Waves

When an earthquake shakes the Earth's surface, vibrations of two basic types travel throughout the Earth.

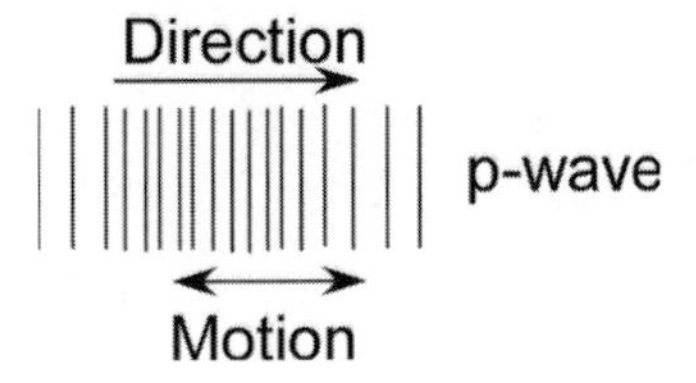

Pressure waves or P-waves move material back and forth in the direction that the wave is moving. These waves can travel through either a solid or a liquid.

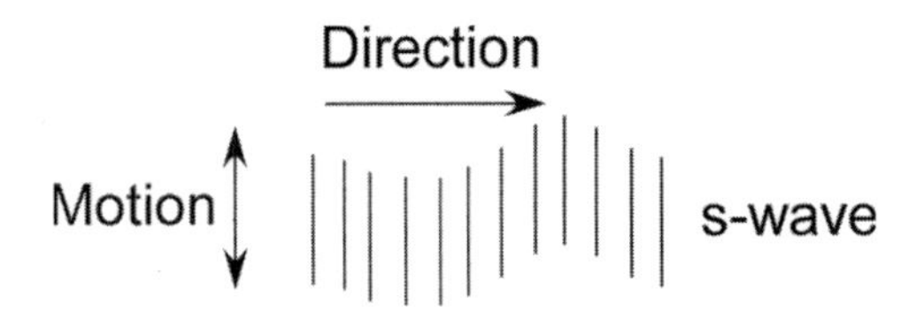

Shear waves or S-waves move material perpendicular to the direction that the wave is moving. If solid rock is sheared sideways like this, it will spring back, so shear waves can travel through solid rock. However, a liquid will not spring back, so these waves do not travel through a liquid.

S waves travel at about half the speed of P waves. Think of P as standing for "Primary" because they arrive first. Think of S as standing for "secondary" because they arrive later.

Probing the Earth's Interior with Sound

When an earthquake occurs, the s-waves and p-waves travel at different speeds and arrive at seismic stations at different times.

When the path between the earthquake and the seismic station is blocked by a region of liquid, the s-wave will not arrive at all. It is said to be "shadowed".

By analyzing the signals received at seismic stations all over the earth, it is possible to figure out the properties of the material that the waves have traveled through.

The process of picturing the Earth's interior by analyzing seismic waves is very similar to the familiar process of using ultrasound imaging to picture an infant in its mother's womb.

- A sudden change in the speed of seismic waves at a depth of about 20 kilometers indicated a sudden change in the nature of the rock at that point, called the Mohorovicic Discontinuity, or *Moho* for short. It marks the transition from granite of the Earth's Crust to the material of the Earth's mantle.

- The shadowing of s-waves indicates a partly liquid core inside the Earth.

- Reflections from the boundary between the Earth's solid core and its liquid core provide a picture of the size of the solid core.

015 Spot Check

Here are some questions to check your understanding of the material in module 015. Both the answers and where to find these questions at the website may found at the end of the Study Guide.

1 The Earth's mantle is made of

a. granite.

b. semiliquid rock.

c. liquid iron.

d. solid iron.

2 The possibility that increasing the amount of carbon dioxide in the air will raise the average temperature of the Earth is referred to as the

a. creation of the ionosphere.

b. creation of smog.

c. Greenhouse Effect.

d. Stark Effect.

e. destruction of the ozone layer.

3 In the Earth's atmosphere, the amount of carbon dioxide is

a. much more than the typical amount of water.

b. about the same as the typical amount of water.

c. much less than the typical amount of water.

4 Pressure waves are transmitted through

a. liquids but not solids.

b. solids but not liquids.

c. both solids and liquids.

5 Colder air always

a. goes westward.

b. goes eastward.

c. rises.

d. moves in circles.

e. sinks.

6 The Ozone layer is where one finds

a. atoms with missing electrons.

b. hurricanes.

c. absorption of ultraviolet radiation.

d. smoke and dust lingering for years.

016: Earth's Living Surface

016.1 An Active Crust

The thin crust of the Earth is one of the least stable solid surfaces in the Solar System.

It has split into large plates that move independently, crashing into each other and sliding past one another.

Evidence from Sea-floor Spreading

The motion of continents was suspected for a long time because of the way they fit together like the parts of a jigsaw puzzle.

The key piece of evidence that Europe and North America are moving apart was the discovery that new sea-floor is being created along the mid-Atlantic ridge.

Europe sits on one moving plate while North America sits on another. They are moving apart by about 2 cm per year.

What Drives the Plates?

The Earth is warmest at the center and cooler farther out: An unstable situation because the warm material wants to rise towards the surface.

The Earth's mantle flows under steady driving forces. Convection cells carry hot mantle material towards the surface while colder material sinks back into the interior. Tectonic plates ride on top of these convection currents.

Sea-floor spreading, such as happens under the Atlantic Ocean, is the result of a rising convection cell in the mantle.

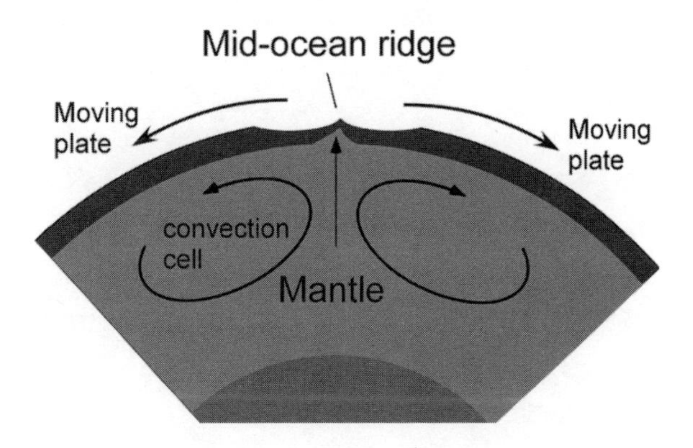

Where there is a descending convection cell, plates are pulled together and either rise up to build mountain ranges or else one plates dives under the other to form a deep ocean trench (as in the Pacific Ocean).

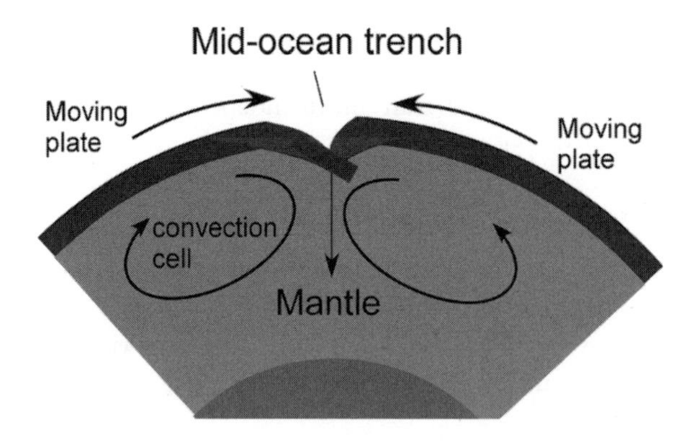

016.2 Earthquakes

Plate Boundary Encounters

When moving plates slip past one another, they stick and slip.

While the plates are sticking or locking together, they become increasingly distorted and build up stored energy.

When the plates finally slip, there is a sudden motion of the earth's crust and all of the stored energy is released as an earthquake.

Evidence from Earthquake Epicenters

The edges of the moving plates are clearly visible in this map of the places where earthquakes have originated.

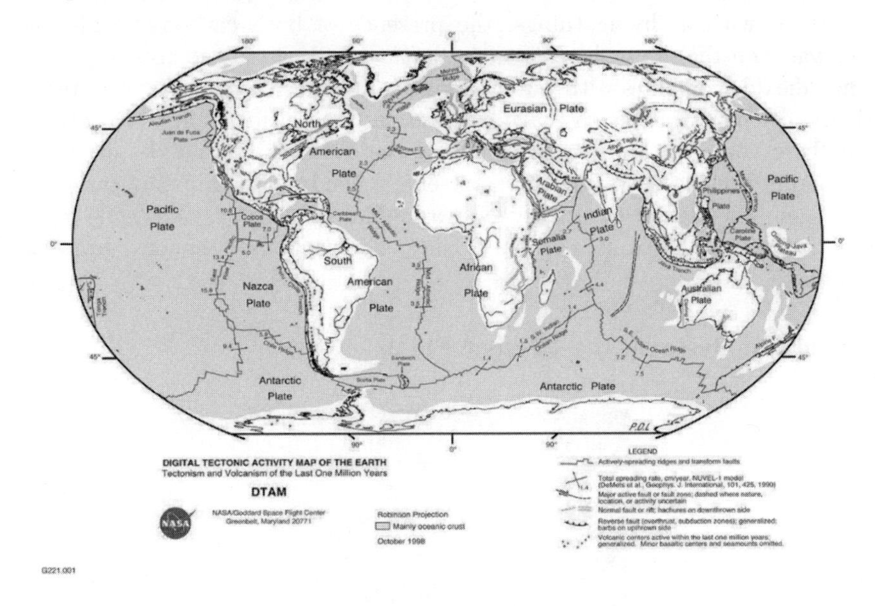

The San Andreas Fault

The state of California sits partly on the Pacific Plate and partly on the North American Plate. These plates meet at the San Andreas Fault.

016.3 The Carbon Cycle

The Unstable Chemical Basis for Life

All life that we know about depends on the presence of *carbon* and *liquid water.* There does not seem to be any substitute for carbon as a basic building block because it forms far more complex compounds than any other element. Here on Earth, carbon first gets into living things in the form of Carbon Dioxide from the atmosphere. Plants use that, together with water, to construct the hydrocarbons that make up their cells.

Likewise, liquid water provides one of the best solvents in nature. By dissolving things, it brings them together so that chemical reactions can take place. Water also provides the hydrogen and oxygen that are needed to make complex compounds with carbon.

With or without living things, the presence of both carbon dioxide and liquid water on the same planet is an *unstable situation.* In the absence of life, Carbon dioxide combines with water to make a weak acid which, in turn, reacts with the calcium in rocks to make calcium carbonate(limestone). Eventually all of the carbon dioxide is washed out of the atmosphere and the carbon is locked up in the rocks. When life is present, it tends to lock up the carbon in organic compounds that eventually get washed into the oceans and settle to the ocean floor. In either case, carbon dioxide is continuously removed from the atmosphere.

The Carbon Cycle on the Present Earth

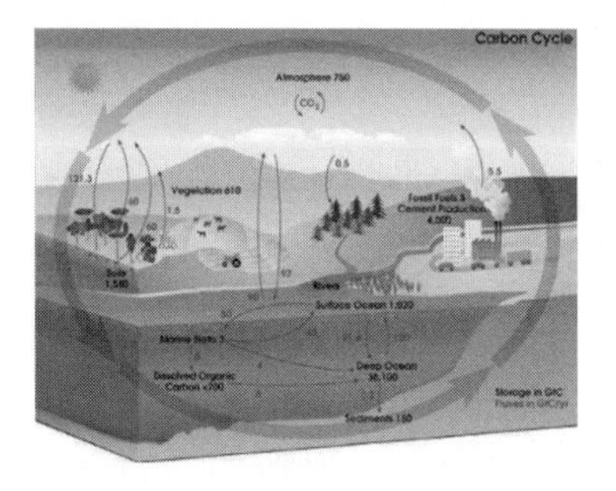

In this picture, the amounts of carbon that move along each branch of the cycle are given in Gigatons (Billions of tons) per year. Notice that all of the different paths lead to the sea floor. The picture shows the carbon getting from there back into the atmosphere, but does not say how it happens.

016.4 Closing the Carbon Cycle

Subduction Zones

Here is the Cascadia Subduction Zone in the American North West:

The zone runs from Vancouver Island to Northern California. A small tectonic plate, called the Juan de Fuca plate is being shoved beneath the North American plate. Extremely large earthquakes and volcanos are common near these zones. The Cascadia zone is thought to be capable of causing earthquakes above 9 on the Richter Scale.

The last known Cascadia quake was in 1700. There appear to have been seven quakes in the last 3500 years.

Volcanos

Subduction zones such as the Cascadia Zone send both limestone deposits and sea-floor sediments deep into the Earth's mantle where they are heated enough to drive off the carbon as carbon dioxide. When the liquid rock or magma from the mantle finds its way to the surface, we have volcanic eruptions and the carbon dioxide that is dissolved in the magma is released into the atmosphere to begin the carbon cycle again.

Although volcanos are an essential part of the carbon cycle that makes life possible on Earth, they can be hard to live next to. The volcanos in the American Northwest are driven by the Cascadia Subduction Zone. These have erupted many times in the past and will erupt again.

016.5 Comparing Earth to Other Planets

Venus

Venus appears to have a thick crust and no tectonic plate motion at all. Since Venus is a near-twin of the Earth, the reason for its lack of moving tectonic plates is not obvious. One idea is that the almost complete lack of water on Venus may have something to do with it.

NASA/JPL

Venus is close enough to the Sun so that any water that it received from the rain of comets would have vaporized. Over time, water vapor that drifted into the upper atmosphere would be taken apart by ultraviolet light from the Sun with the hydrogen being lost and the Oxygen combining with other elements to become part of the crust.

The lack of water meant that there was no mechanism to take carbon dioxide out of the atmosphere, so it stayed there, causing a massive greenhouse effect that heated the planet further, thus driving off any water that might have remained. The carbon cycle never had a chance to start.

It has been estimated that, on Earth, the amount of carbon trapped in sediments because of the action of water is about 250,000 times the amount in the atmosphere. Without that trapping or sequestration of carbon, our atmosphere would be much like that of Venus, almost all carbon dioxide at a pressure 90 times what it is now.

Mars

Mars is only half the size of the Earth and cooled quickly so that it developed a crust too thick to allow tectonic plate motion. The thickness of its crust can be gauged from the size of its mountains such as Olympus Mons

NASA/JPL

which is the largest volcano in the solar system, 24 kilometers high.

Mars does have both carbon dioxide and water. However, the atmospheric pressure is suspiciously close to the triple point of water. That suggests that whenever the pressure rises high enough for liquid water to begin removing carbon dioxide from the atmosphere, the pressure drops below the triple point and the liquid water evaporates or freezes. Here also, the carbon cycle never had a chance to start.

016 Spot Check

Here are some questions to check your understanding of the material in module 016. Both the answers and where to find these questions at the website may found at the end of the Study Guide.

1 Venus retains a dense carbon dioxide atmosphere because

a. there is no life there.

b. there is no liquid water there.

c. there is no plate tectonic activity in its crust.

2 On Earth, the carbon that ends up as ocean sediment and limestone is

a. completely lost to the environment.

b. returned to the atmosphere as methane when the material decays.

c. returned to the atmosphere by volcanos when the sea floor is pulled deep into the Earth.

3 Underneath a place where the sea floor is disappearing into a deep ocean trench, one expects there to be

a. a horizontal current in the Earth's mantle.

b. a magnetic domain in the Earth's core.

c. a descending convection current in the Earth's mantle.

d. a bubble in the Earth's mantle.

e. a rising convection current in the Earth's mantle.

4 Earthquakes are often caused by

a. drought.

b. high winds.

c. slipping tectonic plates.

d. torrential rains.

e. collapsing mountains.

017: Earth Impacts

017.1 Near Earth Objects

Vermin of the Skies

Before the 1970s, astronomers referred to asteroids as the "vermin of the skies" and disregarded their images.

Digitized Sky Survey

An example of one of the 'vermin of the sky': The image was taken in 1955 and is part of the Palomar Sky Survey. It shows the asteroid *1999an10* many years before it was actually discovered.

At that time, estimates of the number of near-Earth asteroids were in the hundreds, mainly because it was thought that such an asteroid would not last very long without hitting something or being kicked into a different orbit.

An Endless Supply

The early estimates were very far off because it was not realized that the supply of near-Earth asteroids is constantly being replenished by new arrivals from the Asteroid Belt.

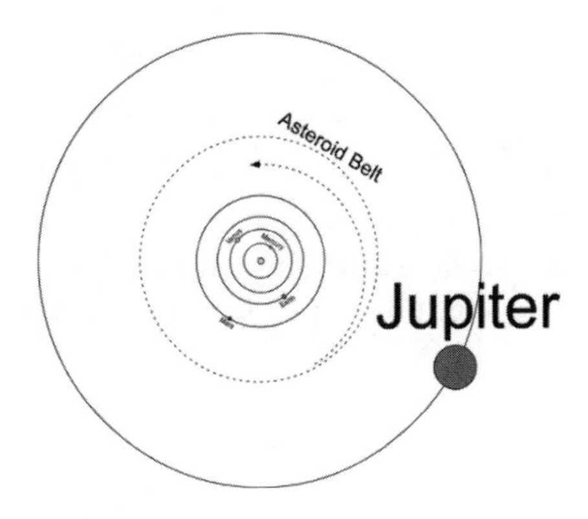

The Belt is constantly stirred by the gravity of Jupiter and other planets.

Asteroids are continually cascading from the Belt into the inner Solar System, where they can cross the Earth's orbit.

Estimated Numbers

Current estimates of the number of near-Earth asteroids (based both on theoretical models and direct sampling. From the House Committee on Science and Technology Subcommittee on Space and Aeronautics Hearing on NEO Survey Programs, Nov. 8, 2007.):

- Over 1 kilometer in diameter: < 1000 (about 850 known in 2007).
- Over 140 meters in diameter: 20,000 (about 5000 known in 2007)
- Over 50 meters in diameter: 200,000 (mostly not trackable in 2007).

These estimates keep changing. At our website you will find, courtesy of the Armagh Observatory, an illustration of how our picture of the inner solar system has evolved from 1800 to 1999 and a current map of objects near us in the Solar System.

The main reason so many asteroids have been discovered since 1990 is the same as the reason for all of the Trans Neptunian Object discoveries: Digital imaging instead of film in telescopes. Asteroids are detected by their motion relative to the distant stars in much the same way as TNOs. Just as for the TNOs, computers are very good at sorting through thousands of pictures and finding the one object that moves.

Many of the TNOs that we discussed earlier were discovered as byproducts of the search for near-earth asteroids.

017.2 Small Object Impacts

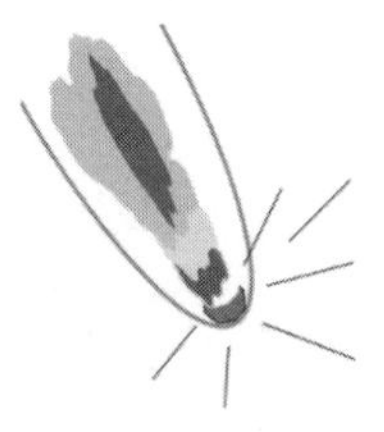

Even asteroids too small to be easily detected before they hit can do a lot of damage. A 50 meter diameter object would release as much energy as a large thermonuclear weapon.

Air Bursts

Some asteroids are made of loose gravel and disintegrate when they hit the atmosphere. However, that is not good news. An airburst releases all of the kinetic energy of the asteroid at once in an explosion capable of devastating a wide area.

In 1908, what is now believed to be a small (less than 50 meter diameter) asteroid exploded in the air about 5 miles above Tunguska, Siberia. The area is so remote that outside observers did not survey the scene of the blast until 1921.

They found trees blown down over thousands of square miles by an explosion that was estimated to be equivalent to a 15 megaton thermonuclear weapon.

From the Leonid Kulik Expedition, 1927

That estimate was made by a crude process of comparing the damage to what would be expected from an air burst of a nuclear weapon. More detailed simulations take into account the fact that the object was descending at more than five miles per second when it exploded and suggest that the actual energy yield was 3 or 5 megatons. That would correspond to a smaller asteroid with perhaps a 20 meter diameter. Unfortunately, that result means that this type of event is more common than we thought and difficult to predict since such small objects are difficult to track.

Chances for Dangerous Mistakes

On June 6, 2002, while India and Pakistan were threatening each other with the use of nuclear weapons, a 12 kiloton blast was detected over the Mediterranean. It is thought to have been caused by an object between 2m and 5m in diameter.

The same blast a few hours earlier would, because of the rotation of the earth, have occurred over land in the disputed region. A nuclear exchange could have been the result.

U.S. Early Warning satellites have been detecting similar explosions with some regularity. A 100 kiloton blast was reported over Greenland in 1996.

(Congressional testimony by Brig. Gen. Simon P. Worden, U.S. Space Command's deputy director for operations at Peterson Air Force Base, Colo., July 10, 2002)

Ground Impacts

Some asteroids are made of iron and hold together until they hit the ground. Their kinetic energy then vaporizes the rock and creates a large crater. Many such craters have been found on Earth.

D. Roddy (U.S. Geological Survey), Lunar and Planetary Institute

The Barringer Meteor Crater in the Arizona desert is a classic example of what happens when a 50 meter diameter iron asteroid hits the ground. The crater is 1.2 kilometers in diameter. It was equivalent to a 15 megaton blast, this time on the ground. The event was 50,000 years ago.

017.3 Large Object Impacts

Impact energy is proportional to mass which is proportional to volume. An object 1km in diameter has 8000 times the volume of a Tunguska or Barringer object and will deliver about 8000 times the energy of those objects - the equivalent of setting off 8000 large thermonuclear bombs.

Asteroid Winter and the Dinosaurs

Such a massive energy release injects smoke and dust into the Earth's stratosphere, where it can stay and block the sunlight for years. During this "asteroid winter," plants die off as do the large animals that depend on them for food.

While exploring for oil in the Gulf of Mexico, evidence of a massive crater was found on the Yucatan Peninsula.

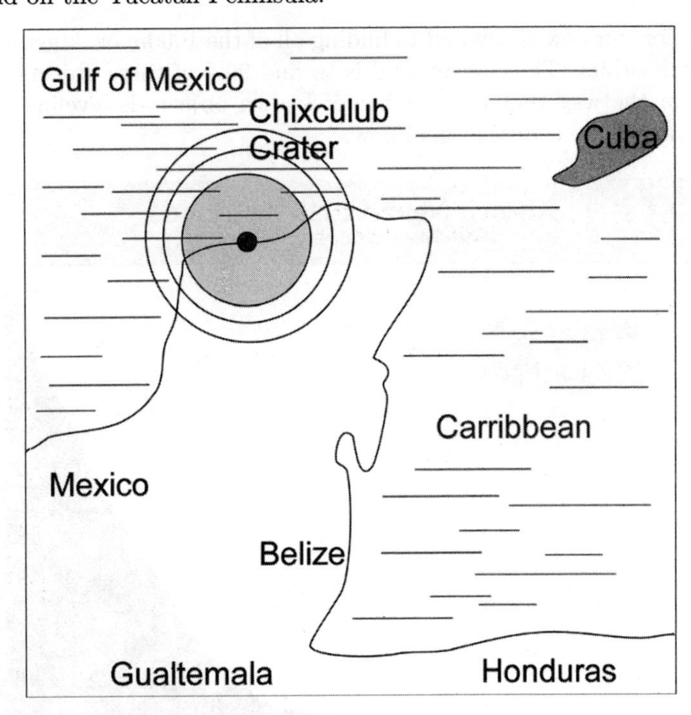

The picture at our website is actually superimposed on a satellite photo of a smoke-covered peninsula. The crater is completely buried and cannot actually be seen from space. Only the folded underground rock strata from the blast betray its presence.

The crater is 300 kilometers in diameter and is thought to have been caused by the impact of an asteroid 10,000 meters in diameter. The energy of impact would have been equivalent to roughly 8 million thermonuclear weapons set off at once.

The Chicxulub impact occurred 65 million years ago and exactly coincides with the extinction of the dinosaurs. That impact wiped out 80% of the species on Earth - every animal larger than a cat and most plant-eating insects.

It has been suggested that this was the event that gave mammals, and eventually humans their chance to dominate the Earth.

What is our risk?

Fortunately, objects capable of causing an extinction-level event are the easiest ones to spot and also have the most predictable orbits. NASA's Spaceguard Program has achieved its goal of identifying at least 90% of near-Earth objects that are larger than one kilometer in diameter.

So far, we are safe. *None* of the objects observed so far poses a threat.

017.4 Hunting Killer Asteroids

The Hunt So Far

The main effort for now is devoted to finding all of the 1000m or larger asteroids in near-Earth orbits. The current goal is to find 90% of these objects. As you can see from the way that the number of known objects is leveling off, it is plausible that we are near that goal now.

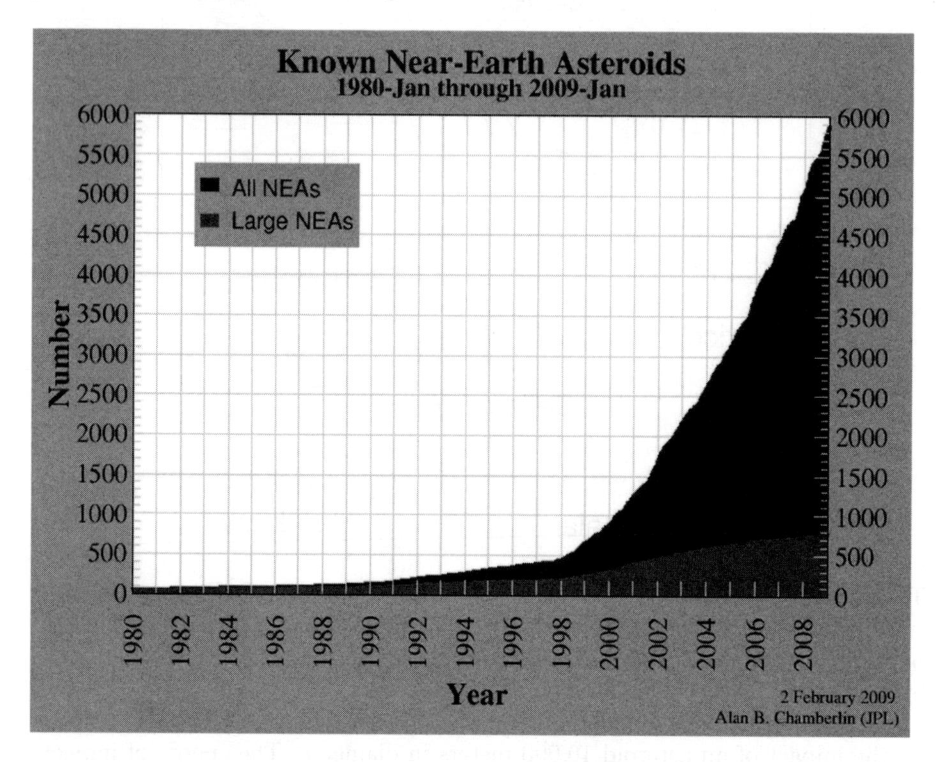

However the detection rate for the smaller objects, like the ones that caused the Tunguska blast and the Barringer Meteor Crater, shows no sign of leveling off anytime soon.

Predicting Asteroid Paths

Because these objects come near the Earth many times, their orbits can be determined precisely.

Asteroids are in free fall, with no random thrusts from non-gravitational forces. Thus, their paths can be predicted for many years in advance.

What happens when an astronomer finds an asteroid on a collision course with Earth?

In June, 1999, an international conference on near-Earth objects was held in Turin Italy. The conference participants voted to use the following "Torino Scale" to describe threats from these objects.

THE TORINO SCALE

Assessing Asteroid/Comet Impact Predictions

Category	Level	Description
No Hazard	0	The likelihood of collision is zero, or is so low as to be effectively zero. Also applies to small objects such as meteors and bolides that burn up in the atmosphere as well as infrequent meteorite falls that rarely cause damage.
Normal	1	A routine discovery in which a pass near the Earth is predicted that poses no unusual level of danger. Current calculations show the chance of collision is extremely unlikely with no cause for public attention or public concern. New telescopic observations very likely will lead to re-assignment to Level 0.
Meriting Attention by Astronomers	2	A discovery, which may become routine with expanded searches, of an object making a somewhat close but not highly unusual pass near the Earth. While meriting attention by astronomers, there is no cause for public attention or public concern as an actual collision is very unlikely. New telescopic observations very likely will lead to re-assignment to Level 0.
	3	A close encounter, meriting attention by astronomers. Current calculations give a 1% or greater chance of collision capable of localized destruction. Most likely, new telescopic observations will lead to re-assignment to Level 0. Attention by the public and by public officials is merited if the encounter is less than a decade away.
	4	A close encounter, meriting attention by astronomers. Current calculations give a 1% or greater chance of collision capable of regional devastation. Most likely, new telescopic observations will lead to re-assignment to Level 0. Attention by the public and by public officials is merited if the encounter is less than a decade away.
Threatening	5	A close encounter posing a serious, but still uncertain threat of regional devastation. Critical attention by astronomers is needed to determine conclusively whether or not a collision will occur. If the encounter is less than a decade away, governmental contingency planning may be warranted.
	6	A close encounter by a large object posing a serious, but still uncertain threat of a global catastrophe. Critical attention by astronomers is needed to determine conclusively whether or not a collision will occur. If the encounter is less than three decades away, governmental contingency planning may be warranted.
	7	A very close encounter by a large object, which if occurring this century, poses an unprecedented but still uncertain threat of a global catastrophe. For such a threat in this century, international contingency planning is warranted, especially to determine urgently and conclusively whether or not a collision will occur.
Certain Collisions	8	A collision is certain, capable of causing localized destruction for an impact over land or possibly a tsunami if close offshore. Such events occur on average between once per 50 years and once per several 1000 years.
	9	A collision is certain, capable of causing unprecedented regional devastation for a land impact or the threat of a major tsunami for an ocean impact. Such events occur on average between once per 10,000 years and once per 100,000 years.
	10	A collision is certain, capable of causing a global climatic catastrophe that may threaten the future of civilization as we know it, whether impacting land or ocean. Such events occur on average once per 100,000 years, or less often.

Fig. 2. Public description for the Torino Scale, revised from Binzel (2000) to better describe the attention or response that is merited for each category.

For example, initial observations of the 250 meter diameter asteroid *99942Apophis* in 2004 indicated a 2.7% chance that it would hit the Earth in 2029. Additional observations refined the orbit and showed that it would miss the Earth in 2029. However there was still a small region in space near the Earth, called a *gravitational keyhole* just a few hundred meters across. It remained possible that Apophis might pass through that keyhole, setting up an Earth impact on April 13, 2036. For some time Apophis was rated at level 1 on the Torino Scale and, at one point got up to level 4, which so far is an all-time record.

Additional observations of Apophis, using radar as well as newly identified previous observations, have refined the orbit prediction even more and the impact probability for 2036 is now estimated to be 1 in 45,000. An additional possible impact date in 2037 has been found, but its probability is only 1 in 12.3 million. As a result, Apophis has now been dropped to 0 on the Torino Scale.

Risk Assessment for Rare Events

Science can provide basically *two measures* for these impact events, the *probability of impact*, and the *consequences of an impact*. It is then up to us to decide when those two measures warrant taking some sort of action.

For Apophis the probabilities are now estimated to be small. However the severity would correspond to the explosion of 880 million tons of TNT, equivalent to several of our larger thermonuclear weapons going off at once. That is not enough to cause an asteroid winter, but it would be far worse than the Tunguska event and a very bad day for anyone near the impact.

It is possible to predict a most likely impact along a path stretching from southern Russia, across the Pacific (near the California coast) and then down into Colombia and Venezuela. Impact near California would cause deadly tsunamis with casualties that are difficult to predict. An impact in Colombia or Venezuela is estimated to cause more than 10 million casualties.

The Mathematics of Death

One way to measure the severity of an event is to work out the long-term average death rate. The impact probability of 1 in 45,000 means that if we let 45,000 Apophis-type asteroids go undisturbed, we would get hit once. Suppose that a hit would kill 10 million people. The average death rate is then 10,000,000 divided by 45,000 events or 222 people per event. On a planet with over 6 billion people, that is a pretty small number compared to many other dangers such as automobile accidents and insect bites and would not seem to be worth worrying about.

However, this sort of long-term average rate is not what we normally respond to. Insurance companies make money because people will pay *much* more than their probable average losses to avoid a catastrophic loss. There is a *big* difference between an average casualty rate of 222 per event and a small but non-zero chance of killing 10 million people.

One proposal, called 'Operation Foresight' is to spend approximately 137 million dollars to launch a simple space probe to shadow Apophis and refine its orbit to the point of either ruling out impact altogether or predicting a near certainty of impact. Is that a poor bargain for keeping a statistical average of 222 people alive or is it cheap insurance against a small but real chance of 10 million casualties?

A cold and rather heartless way to do the math is to figure out the lost productivity of the people killed. In a developed economy, a premature death can easily mean about a million dollars in lost earnings. In those terms, Operation Foresight might be justified since it would spend 137 million dollars to avoid an average loss to society of over 222 million dollars.

What happens when the downside risk is not just the loss of a major city or two but the end of our species? How do we do the math then? In that case, the lost productivity would not just be that of the six billion or so people that would be killed. It would also be the productivity of all of the descendents that they would never have. In that case, it is easy to argue that even a very tiny probability of such an impact would justify whatever it would cost to prevent it.

017.5 Asteroid Defense

Shoot it Down: A Bad Idea

The main problem with a Star Wars approach to asteroid defense is that we may not know enough about the detailed composition of a given asteroid to predict how it will respond to a violent attack.

A solid rock asteroid might be shattered into pieces that remain large enough to impact the ground. In that case, the amount of energy and thus the damage that is delivered to the surface would not be changed very much.

Some asteroids are little more than collections of loose gravel held together by their weak gravitational attraction. A nuclear explosion might just make the rubble bounce a bit and then pull itself back together so that nothing is accomplished.

Land on it and Push: Maybe Not

Space probes have landed on asteroids before. One idea is to land and attach some sort of rocket to the asteroid and shove it off course.

The main difficulty with this idea is the same as for the Star Wars approach: We will not necessarily know the structure of the asteroid and how it will respond to being pushed at just one point. If the asteroid is actually a loose collection of rocks, we might end up deflecting just one rock away from the main rock-pile.

Another difficulty is that a solid asteroid of the sort that would lend itself to this approach will probably be spinning, which would complicate landing on it.

The Gravitational Tractor: Maybe

A recent idea is to send a space probe on a course parallel to the asteroid and have it maintain its position so that the gravitational attraction between the probe and the asteroid gradually pulls the asteroid off course. A 20 ton nuclear-powered spacecraft of this sort could deflect a 200 meter diameter rock and iron asteroid by hovering near it for about a year.

A big advantage of this approach is that it does not depend on the structure or composition of the asteroid. Another advantage is that it can be combined with the Operation Foresight idea so that the probe could just shadow the asteroid for a while to get an absolutely accurate prediction of its path and then close in on it if it needs to deflect it.

A disadvantage is that the probe needs to have a lot of mass and a long-term power source and that is expensive.

017 Spot Check

Here are some questions to check your understanding of the material in module 017. Both the answers and where to find these questions at the website may found at the end of the Study Guide.

1 The number of near-Earth asteroids is large because they

a. are in stable orbits and have nowhere else to go.

b. are left over from the formation of our Moon.

c. are kicked out of the asteroid belt by Jupiter's gravity.

d. are the remains of a destroyed planet near the Earth.

2 You hear about an asteroid impact threat at level 10 on the Torino Scale. You should

a. suspect an error because the scale does not go that high.

b. forget about it, nothing is going to happen.

c. hang loose until the situation becomes clearer.

d. think about ways to get off the planet.

3 An asteroid impact that leaves a huge crater is probably due to an asteroid that is made of

a. frozen gas and ice.

b. gold.

c. iron and nickel.

d. rocks loosely held together.

4 A large asteroid impact can affect the Earth's climate primarily by

a. blocking the sunlight with its smoke and dust.

b. producing strong winds from its passage.

c. poisoning the air with its fumes.

d. heating the air with the heat of impact.

018: Requirements for Life

0.1 018.1 Light

Light is a requirement for most life on Earth and is also the messenger that brings all of our information about the rest of the universe. Before continuing our discussion, we need to know a little about it.

Frequency and Color

Light is an electromagnetic wave of the same type as radio waves. Like a radio wave, it can be described by its frequency. Here is a quick summary of all of the different frequency ranges of electromagnetic radiation:

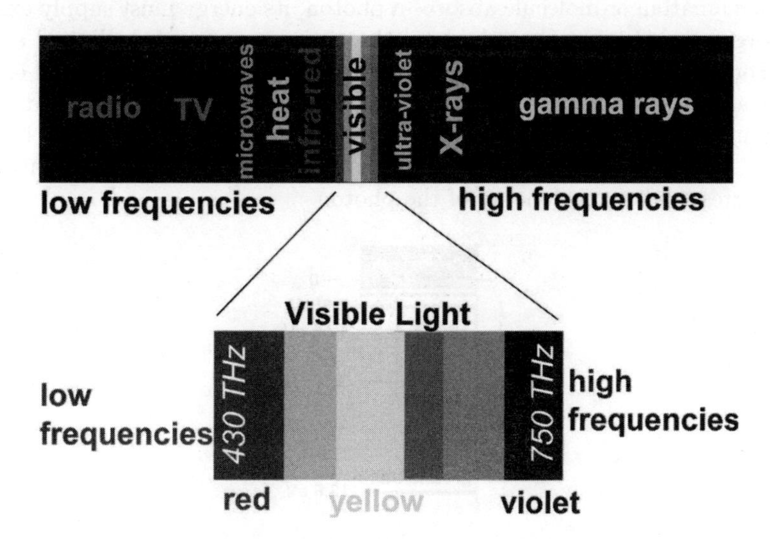

The range of frequencies that corresponds to visible light is rather narrow compared to the enormous range of frequencies for all the types of electromagnetic radiation.

Within the visible range, different frequencies of light correspond to different colors. For example, an electromagnetic wave that is oscillating at 750 THz (7.5×10^{14} cycles per second) would be seen as violet light while a wave oscillating at 566 THz would be seen as green light and a wave oscillating at 430 THz would be seen as red light.

Photons

Light has two different descriptions: It is a wave and it is also made up of particles called *photons*. The two descriptions are connected by the formula

$$E_{\text{photon}} = hf$$

where E_{photon} is the energy of a single photon, f is the frequency of the wave and h is a constant, called Planck's constant. Thus:

- high frequency light corresponds to high energy photons and
- low frequency light corresponds to low energy photons.

Emission and Absorption of Light

Light can only be emitted or absorbed as photons. In addition, atoms and molecules have only certain allowed energy levels. These two facts make it possible for astronomers to know what elements and compounds are present in distant stars and galaxies.

When an atom or molecule absorbs a photon, its energy must supply exactly the energy needed to go from the level that it is on to another allowed energy level. For each element or compound there is a distinctive set of allowed energy levels and thus a distinctive set of photon energies that can be absorbed.

For hydrogen, the allowed absorption of visible light is described by this diagram. Each arrow represents a photon being absorbed. The length of the arrow corresponds to the energy of the photon.

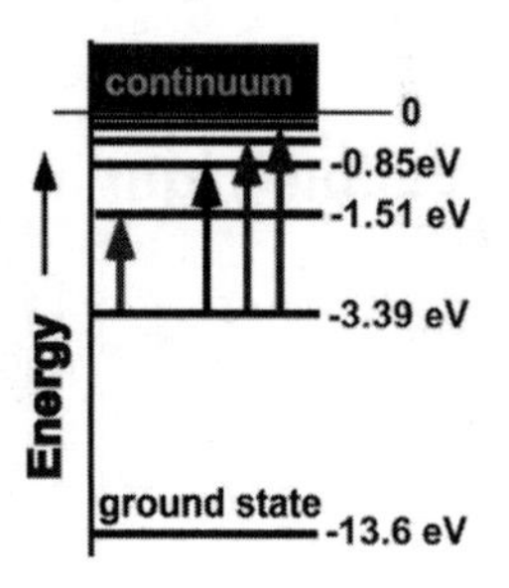

Because the photon energy just corresponds to the frequency of light, each element or compound absorbs a distinctive set of frequencies from any light that passes through it. In this way, each chemical substance leaves a distinctive imprint on the light that passes through it. Astronomers have only to spread the different frequencies of light out to see which ones are missing.

When the light from a star (such as our Sun) is spread out into a rainbow of colors, some colors are missing. This kind of picture is called a *spectrum.* The missing colors in such a spectrum correspond to light that has been absorbed in passing through the outer atmosphere of the star. The pattern of missing colors makes it possible to tell what elements are present and the relative amounts of those elements.

018.2 The Chemical Basis of Life

Abundances

Any form of life must be made from the 90 stable chemical elements that are available in nature. As we just described, these chemical elements leave their signatures in the light that comes to us from stars. By analyzing the relative abundances of elements in the stars that we can observe, we can arrive at the relative abundances of elements in the universe.

Compare these abundances (by mass) to the corresponding abundances in a human body:

Element	**Universe**	**Human Body**
Hydrogen	73.9%	10.0%
Helium	24.0%	0.0%
Oxygen	0.01070%	65.0%
Carbon	0.0046%	18.0%
Neon	0.00134%	0.0%
Iron	0.00109%	< 0.05%
Nitrogen	0.00095%	3.0%
Silicon	0.00065%	0.0%
Magnesium	0.00058%	0.05%
Sulfur	0.00044%	0.20%
All Others	**0.00065%**	→

All Others	
Element	**Human Body**
Calcium	1.50%
Phosphorus	1.20%
Potassium	0.20%
Chlorine	0.20%
Sodium	0.10%
Cobalt	<0.05%
Copper	<0.05%
Zinc	<0.05%
Selenium	<0.01%
Fluorine	<0.01%

Notice that 93% of the human body is made up of carbon, hydrogen, and oxygen. These are *three of the four most common elements in the universe* and the one that is left out, Helium, is a noble gas that does not participate in chemical reactions.

Also notice that we do require significant amounts of a few relatively uncommon elements (calcium, phosphorus, potassium, chlorine and sodium).

Why Carbon-based Chemistry?

Carbon-based life is already made of the most common elements in the universe. Life based on a different set of elements would be using much scarcer materials and would be much less likely.

After several hundred years of chemistry, we have yet to find any element that forms a variety of compounds that is anywhere close to the ones formed by carbon. With four chemical bonds and the ability to bond strongly to itself, carbon is the ideal backbone element for extremely complex molecules.

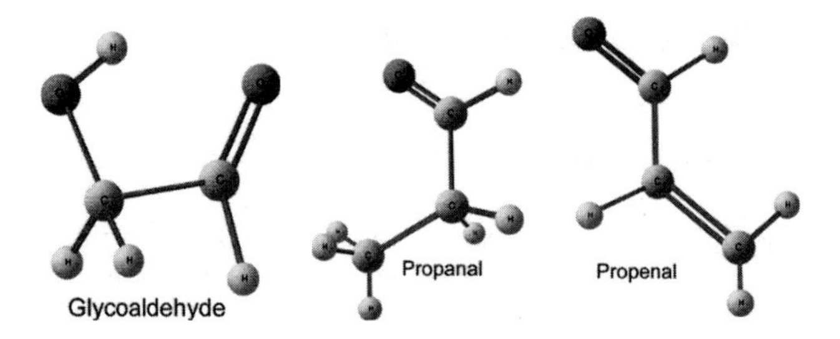

Notice how the carbon atoms form the backbones of these molecules. Also notice the double bond between carbon atoms in the propenal molecule.

What About Silicon-based Life?

One proposal for an alternative chemistry for life is that carbon might be replaced by silicon. Silicon is less common than carbon, but not by a large factor and it is extremely common in the crusts of terrestrial planets. For example, it is about 25% of the Earth's crust.

Silicon, like carbon, can form four chemical bonds and it does form large molecules (called silanes) that are analogous to the hydrocarbons that make up carbon-based life forms. The difficulty is that the silicon-silicon bond that is needed to make up the backbones of these molecules is *much weaker* than the carbon-carbon bond and the silanes are mostly unstable. In fact, they tend to explode in the presence of oxygen. (If you ever meet a silicon-based alien here on Earth, do NOT let it open its spacesuit!)

A related problem is that silicon atoms do not form double and triple bonds with themselves (like in the propenal molecule above) and that reduces the possible variety of molecules.

Where carbon dioxide dissolves readily in water and lends itself to the carbon cycle that transports carbon through the life forms on Earth, silicon dioxide is a solid that does not dissolve in water. The role of water as a solvent to promote chemical reactions is fundamental to Earth-based life and it is very difficult to think of a substitute solvent that would work for a silicon-based biochemistry.

Finally, carbohydrates such as the glycoaldehyde pictured above have been found in interstellar clouds and in meteorites while complex silicon compounds have not.

Life is Metal-rich

Astronomers refer to all of the elements heavier than carbon as "metals" and speak of a star as *"metal-rich"* if its spectrum indicates large amounts of the less common heavier elements such as iron and calcium. Although carbon-based life is mostly made from carbon, oxygen, and hydrogen, it also depends on the special properties of less common elements such as iron and calcium. Those elements would be missing from the planets of a metal-poor star, so we would expect life to be less likely to exist there.

018.3 The Requirements for a Carbon Cycle

Liquid Water

A molecule of water consists of two hydrogen atoms attached to an oxygen atom. In water, each hydrogen atom donates its electron to the oxygen and is left as little more than a naked proton hanging on outside. The oxygen then has a negative electric charge while each of the hydrogens has a positive electric charge. The special properties of water come from the non-symmetrical way that the two hydrogen atoms are arranged.

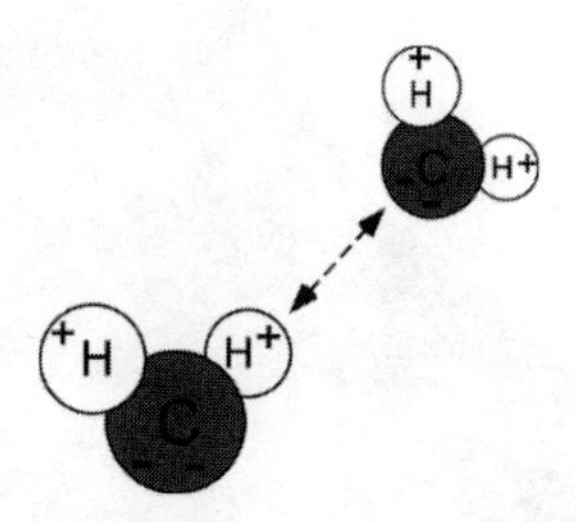

Because opposite electrical charges attract, there is an attraction between the positively charged hydrogen atoms and any negative charge. Similarly, there is an attraction between the negatively charged oxygen atoms and any positive charge. These attractions are called *hydrogen-bonds.*

The hydrogen bonds between water molecules serve to keep water liquid over a very large range of temperature. Think of a group of square dancers who are in constant motion but are always holding the hands of other dancers. At low temperature, the molecules slow down until the hydrogen bonds form a very open crystal with each molecule bonded to just four neighbors. That ice crystal takes up more space than the square-dancing molecules of water in liquid form, so water expands when it freezes. Without that very special property, lakes and oceans would freeze from the bottom up and would mostly stay frozen all year.

Instead, the less dense ice floats on top of the water and provides insulation that keeps the rest from freezing and protects any life forms below.

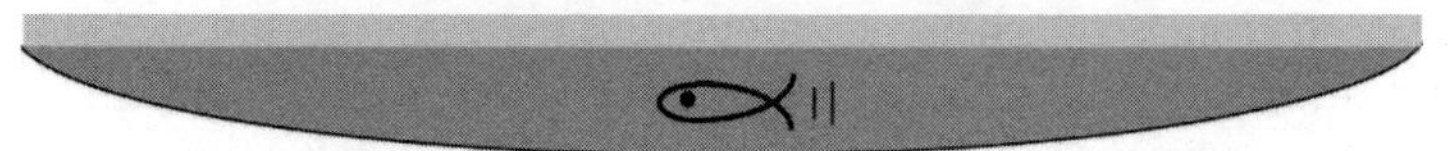

Besides its ability to stay liquid and to freeze in a friendly fashion, water forms hydrogen bonds with many different kinds of molecules, thus holding them in solution and, in many cases, bringing different molecules together so that they can react. All of the processes of life take place in this sort of water-solution environment.

Carbon

Here on Earth, carbon is incorporated into living things as carbon dioxide. Because it is a gas that dissolves in liquid water, it is a particularly convenient and portable source of carbon. However there can be other sources of carbon.

Meteorites have been found to contain complex organic compounds. A notorious example is the Murchison meteorite, which fell near Murchison, Victoria in Australia on Sept. 28, 1969.

http://www.meteorites.com.au

The meteorite was recovered a very short time after it fell and was found to contain amino acids and other complex hydrocarbons. Arguments about whether or not these particular hydrocarbons came from space or are the result of earthly contamination have not yet been settled. However there is no doubt that hydrocarbons do exist in space since we see their spectral signatures in the radio waves from interstellar clouds. The early Earth may have had an abundance of such hydrocarbons to jump-start life and the carbon-cycle.

Other possible sources of carbon, particularly methane, could provide alternatives to carbon dioxide as the portable form of carbon. Titan, for example, is known to have large amounts of methane in its atmosphere.

A Carbon-return Process

To complete the carbon cycle, there must be some way to return all of the carbon to its portable form. Here on Earth the carbon tends to end up in sediment on the ocean floor and combined with calcium in surface rocks. The return process here is plate tectonics which recycles the surface into the hot interior of the planet and returns carbon dioxide to the atmosphere through volcanic action.

Plate tectonics requires a planet that remains semi-molten except for a thin surface crust. A planet that is too small, such as Mars, would cool too quickly so that plate tectonics could never start. It appears that there is a minimum size somewhere between the size of Mars and the size of Earth. Planets smaller than that minimum size might not be able to sustain a carbon cycle.

018.4 The Energy Sources of Life

Energy is Conserved

The key property of energy is that it cannot be created or destroyed. It can, however, be stored and released and transported from one place to another.

For example, a mixture of hydrogen and oxygen will release energy (explosively) if it combines to make water. To reverse the process and separate water into hydrogen and oxygen requires one to put back the same amount of energy that was released when the water formed.

The splitting of water into hydrogen and oxygen is a key step in the process of making complex hydrocarbons from a mixture of water and carbon dioxide. It can only happen with an energy input of some sort.

Energy from the Sun

For most life here on Earth, the basic energy input comes from the Sun.

Plants use a complex molecule-sized chemical factory consisting of several Chlorophyll molecules to convert water into oxygen and hydrogen. The hydrogen is generated as separated protons and electrons which are then used in other processes to convert carbon dioxide into complex hydrocarbons. The oxygen is released into the atmosphere.

Each step of the process that Chlorophyll carries out needs an energy input in the form of photons from the Sun. The largest number of photons that reach the Earth's surface are photons of red light, so Chlorophyll absorbs those and uses their energy. The highest energy photons that reach the Earth's surface are photons of blue or violet light, so Chlorophyll also uses those after stepping their energy down to match its needs. The middle of the visible spectrum has photons that are not the most abundant and are also not the most energetic, so Chlorophyll does not use those. As a result, we have plants that absorb energy from either end of the visible spectrum and reflect the green light that is in the middle.

Energy from Hot Water

Energy is a lot like money. You can buy the same things with it no matter where it comes from. Similarly, life has shown an incredible flexibility in using different sources of energy. Even here on Earth, not all life gets its energy from the Sun or from eating plants that got their energy from the Sun.

One alternate energy source is the heat from the interior of the Earth, the same thing that drives plate tectonics. At sea vents, called *black smokers*, in the floors of both the Atlantic and Pacific oceans, water heated to $400°C$ jets upward into the ocean. The water does not boil because these vents are typically 2100 meters below sea level and the pressure keeps the superheated water in liquid form. The water contains dissolved metal sulfides that precipitate out when they hit cold sea water, forming black chimneys.

NOAA

Each year, about 370 trillion gallons of water passes through these smokers.

Bacteria use a process called *chemosynthesis* to convert methane and carbon dioxide into organic matter by using the oxidation of hydrogen or hydrogen sulfide or ammonia as a source of *chemical energy* instead of using sunlight. The bacteria then support more complex organisms such as tube worms that are filled with symbiotic bacteria. These can grow to as much as eight feet long.

Photosynthesis in the Dark

Photosynthesis does not have to use sunlight. At a depth of 2500 meters off the coast of Mexico there are *green sulfur bacteria* that use the faint glow (mostly invisible infrared light) from a black smoker to power their activities.

018.5 Reproduction

DNA

A defining property of a life-form is that it can make copies of itself. All life on Earth uses exactly the same copying system, based on the DNA molecule. The molecule looks like a twisted ladder with pairs of amino acids forming the rungs and a sugar phosphate polymer forming the sides.

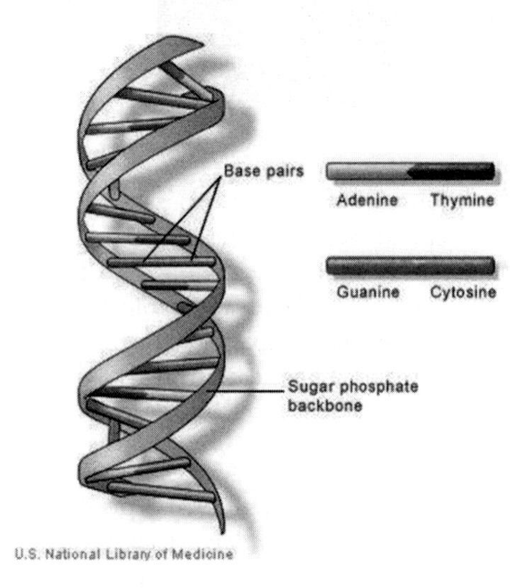

The copying scheme is based on the idea that one long-chain organic molecule can act as a template for forming another one. A DNA molecule acts as a template for making a copy of itself by splitting each amino acid pair to make two half-ladders which then act as patterns to form their missing halves.

Segments of the DNA molecule act as patterns for forming intermediate half-ladder molecules which are the various forms of RNA. Some of the RNA molecules carry out chemical reactions themselves while others act as patterns for proteins and enzymes, the large molecules that are needed to operate an entire living creature.

How Much is Accident?

Every living thing on Earth uses exactly the same DNA/RNA scheme with four amino acids encoding its basic operating system. Furthermore, just 20 amino acids are always used as the basic building blocks of enzymes. There are hundreds of known amino acids, so what is so special about these particular ones?

A revealing point is that each of these molecules has a mirror-image molecule that is identical, but with everything reversed as it would look in a mirror. The mirror-image molecules can carry out exactly the same chemical reactions among themselves as the originals but are never found in living things.

It is thought that all life on Earth is related and is all descended from a single original self-replicating molecule that made all of the apparently arbitrary choices of which amino acids to use.

How (and where) did it start?

There are two main competing models for how life on Earth started: The *Abiogenesis Model* and the *Panspermia Model*. Abiogenesis says that the original self-replicating molecule that led to Earth life formed spontaneously from non-living material here on Earth. The Panspermia Model is based on the discovery that material from impacts on other planets has been deposited on Earth. Because bacterial spores and extremophile bacteria are known to be extremely hardy, they could have survived long space trips and brought life to Earth from elsewhere.

In each model, the first self-replicator had to have formed *somewhere*. The key difference is just *how difficult* that first step really is.

It cannot be too easy or we would expect it to have happened several times here on Earth and some isolated remnants of the less successful versions would still be around.

If it is moderately easy so that life arises just once when the conditions are right for it, then we have the abiogenesis model.

If the formation of the first replicator is actually very difficult, then it may have happened only once in our Solar System or possibly only once in the entire universe. In that case we have the panspermia model.

The two models make very different predictions of what sort of life we might find elsewhere. The abiogenesis model suggests that any life that we find on Mars or on the Moons of the Jovian planets would not have Earth-type DNA and might even be using an entirely different scheme for reproducing itself. The panspermia model suggests that we will find exactly the same DNA and the same biochemical processes everywhere that is within range of whatever process carried the original seeds of life.

018 Spot Check

Here are some questions to check your understanding of the material in module 018. Both the answers and where to find these questions at the website may found at the end of the Study Guide.

1 The key advantage of Carbon over Silicon as the basic element for life is that

a. carbon is more abundant than silicon.

b. carbon atoms can bond to other carbon atoms but silicon atoms cannot bond to other silicon atoms.

c. carbon can combine with hydrogen and silicon cannot.

d. carbon forms more stable compounds than silicon.

e. carbon-based life got started first.

2 In current Earth life, the self-copying molecule that contains the information needed to construct and operate a life-form is

a. glycoaldehyde.

b. chlorophyll.

c. glucose.

d. RNA.

e. DNA.

3 The molecules of both liquid water and ice are held together by

a. transferring electrons between water molecules to make electrically charged ions.

b. the attraction between the hydrogen atoms on one water molecule and the oppositely charged oxygen atom on another.

c. weak electrical fluctuations in one water molecule and the opposite electrical fluctuations that it induces in another.

d. sharing electrons between water molecules.

4 Green plants obtain the energy they need from

a. the soil they are planted in.

b. hot water.

c. sunlight.

5 Which of the following types of radiation has the lowest frequency on this list?

a. microwaves.

b. infrared light.

c. ultraviolet light.

d. red light.

e. X-rays.

6 A spectrum is defined to be the set of

a. frequencies or colors missing from a light source.

b. energy levels missing from a light source.

c. photons present in a light source.

d. energy levels in a light source.

e. frequencies or colors present in a light source.

7 Compared to the frequency of photons absorbed during a transition from a from a -5ev state to a -4ev state, transitions from the -5ev state to a -1ev state would correspond to absorbing photons whose frequency is

a. 4 times as high.

b. 3 times as high.

c. 5 times as high.

d. 2 times as high.

e. the same.

8 An advantage of carbon dioxide as the starting form of carbon for use by living things is that it

a. dissolves in liquid water.

b. is the result of burning carbon.

c. freezes to form dry ice.

019: The Search for Life

019.1 The Motivation

How did life begin?

Besides deciding between the abiogenesis and panspermia models for the origin of life, we would like to find an example of life in a more primitive state than exists on Earth. That would provide some real information about how life here may have started.

Even if the life that we find is far from its primitive state, it may have followed such a different path from Earth life that we can learn something about a possible common starting point.

How does Biology really work?

Current biologists are in the position of a mechanical engineer who has just one single machine to study. The discovery of completely different biochemical systems would change the nature of the discipline and could lead to many useful new insights into our own biochemistry.

Are we alone?

Some scientists think that the first self-replicator was *so extremely improbable* that it formed *only once* in the universe, and that was here on Earth. If they are correct, we will not find life anywhere but here on Earth.

Almost everyone seems to have a strongly negative emotional reaction to this idea. Partly it is our experience with the Copernican Principle, which says that we are never special. We have learned that Earth is not the center of the Solar System and our Sun is a very ordinary star among hundreds of billions of stars. Partly it is a feeling that life is the purpose of the universe and a universe with only one living planet in it seems like a tremendous waste. Finally, the idea of our being absolutely alone in the universe and responsible for all of the life that exists in it is very scary.

Just one genuine fossil microbe from another planet would be enough to put this extremely unpleasant possibility to rest.

019.2 Mars

The Requirements for Life

Mars has an atmosphere of carbon dioxide, so the *carbon source* is there. However there is *no global carbon return process* since Mars does not have any tectonic activity. Any life that exists now would have to have some sort of local carbon return process.

Mars has water ice at its poles and water vapor in its atmosphere. However the atmospheric pressure is near the triple point of water so that liquid water cannot usually exist on the surface.

There is some evidence of *subsurface liquid water* as in this Mars Global Surveyor picture of sudden gullies:

NASA/JPL

It is thought that these occur when subsurface liquid water, about 100 meters down, is brought to the surface by the formation of an ice dam.

Sunlight is a possible *energy source* at the surface, but conditions there are probably too hostile for life to take advantage of it. Subsurface life would have to use some form of chemical energy.

Past Conditions

There is abundant evidence that Mars once had liquid water on its surface, which would imply a much higher atmospheric pressure in the past. If conditions were favorable for a long enough time, life might have developed and we could hope to find microbial fossils left from that time.

If there really is subsurface liquid water it is possible that life may have evolved on the surface and gradually gone underground as conditions worsened.

The Viking Experiments

The Viking landers (1976) were equipped to look for evidence of biological activity in the surface soil. The landers scooped up some Martian soil and placed it in an experiment chamber for analysis by the following experiments:

Gas Exchange Looked for changes in the composition of the gas in the test chamber to indicate respiration. **Result: No activity.**

Labeled Release Fed radioactively tagged nutrients to the soil sample and looked for the tags in the gas from the sample. **Results: Inconsistent.**

Pyrolytic Release Fed radioactively tagged carbon dioxide to the sample and then heated the sample to see if it had been used to make organic compounds. **Results: 7 of 9 runs showed positive results. Later discounted.**

Mass Spectrometer Heated the sample to drive off any water or organic compounds and measured how much was present. **Result: More water than expected and no organic compounds at all.**

The Mass Spectrometer Experiment indicated such an absolute lack of organic compounds that something (probably ultraviolet light from the Sun) had to be actively destroying them.

In order to avoid false positives, the same experiments were performed on a **control sample** that had been heat-sterilized. The control sample gave positive results for the Gas Exchange and the Pyrolytic Release experiments. Those results were later reproduced in Earth laboratories and shown to result from chemically active oxides generated by the intense ultraviolet light at the surface of Mars.

The **final conclusion** was that the surface of Mars is extremely hostile to any form of life.

Martian Meteorites

About 34 meteorites (out of about 24,000 that have been found) appear to have originated on Mars. One way that they are identified is to compare trapped gas bubbles with the Martian atmosphere.

In 1996, the meteorite ALH 84001 caused a great deal of excitement. It was found to contain carbonates (orange material in the picture at our website) that could only form in liquid water. It was dated to about 4.5 billion years ago and contained what NASA scientists claimed were fossils of microorganisms.

Arguments about the nature of these structures continues, with some claims that they are the result of Earth contamination and other claims that they are not of biological origin. One problem with them is that the are much smaller than Earth bacteria and smaller than the generally accepted lower size limit (200 nanometers) for living organisms.

019.3 The Jovian Moons

Titan

Saturn's largest moon, Titan, has an atmosphere that is about 1.6% methane (CH_4) along with lakes of liquid methane and related liquid hydrocarbons, so there is an abundant source of *portable carbon* that could play the role that carbon dioxide does here on Earth.

Observations by the Cassini orbiter indicate that landmarks on the surface of Titan are not rotating with the rest of the moon and are shifting by as much as 19 miles from their expected positions. The shifting is consistent with an ocean of *liquid water* and possible ammonia about 62 miles below the surface.The unstable nature of Titan's surface means that there could be a *carbon return* process similar to plate tectonics on Earth.

If the water or water-ammonia ocean has vents that are being heated from the interior of the moon, chemical *energy* could be available to support life similar to that near deep sea-vents on Earth.

Enceladus

Saturn's moon Enceladus has shown plumes of ice crystals that are believed to be geysers of *water* from pressurized subsurface reservoirs. This picture was taken in February 2005 by Cassini

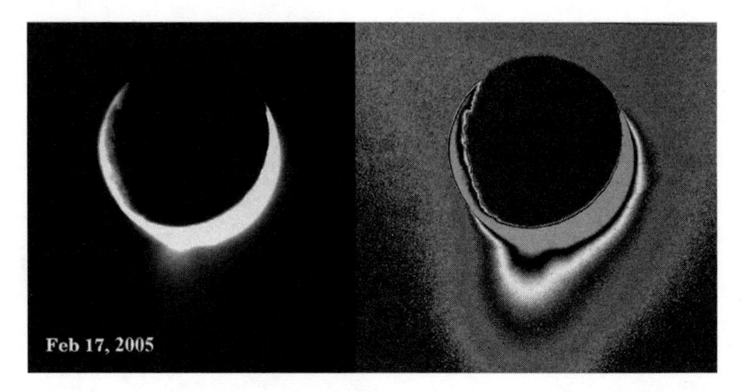

In March 2008, Cassini flew directly through the plume and collected ice crystals from it.The geysers are sending material out at about 800 miles per hour, so there is clearly an abundant *energy source* that is driving them.

Life powered by chemical or heat energy could exist in the subsurface water reservoirs.

Europa

Jupiter's moon Europa was the first of the Jovian moons to be suspected of having a subsurface ocean of *liquid water*. The patterns of cracks in the ice on its surface and the lack of craters there suggest that it is just the frozen top of an ocean. The moon is unusually dense (3000 kg/m^3) for a Jovian moon and is thought to have an iron core similar to the Earth. Its oddly cracked surface with no traces of craters is thought to be mostly frozen water ice.The liquid water interior is kept warm by *tidal energy* generated by varying gravitational forces that stretch the planet. The actual thickness of the ice crust is a subject of dispute and may be as little as a hundred meters but could also be many kilometers thick.

As is the case for Titan and Enceladus, life could exist within the subsurface ocean, powered by tidal energy.

0.1.1 Other Jovian Moons

Subsurface oceans appear to be common among the Jovian moons. Jupiter's moons Ganymede and Callisto are thought to have deeply buried oceans of liquid salt water. Those oceans were detected by the Galileo orbiter from their effects on the magnetic field near them.

In a 2003 report, NASA focussed on Callisto as a possible site for a human base. Like the other Jovian moons, its surface is largely composed of water ice, which can be melted for water and hydrolyzed to make rocket fuel. In addition, it has a stable surface and is located well outside of Jupiter's large radiation belts.

019.4 Extrasolar Planets

Requirements for Earthlike Life

Life of the sort that exists on the *surface* of the Earth requires an atmosphere whose pressure and temperature support the existence of liquid water. For a star such as our Sun, Venus is clearly too close to the Sun to retain any water at all and Mars is clearly too far away. The range of distances where planets can have liquid water is called the *habitable zone.* For our Sun, the habitable zone is thought to be from 0.95au 1.37au.

Notice, by the way, that Earth appears to be at the *inner edge* of the habitable zone — on the verge of being too hot to support life. Since stars normally become more intense with age, we are near the end of our run as a habitable planet. By some estimates, we have only about 500 million years left, assuming that we do not heat things up faster than that by our own activities.

The narrowness of the habitable zone means that usually there will be just one planet in that zone. In our Solar System, we are that planet. To find another planet in a habitable zone, we must look to planets in orbit around other stars.

An additional requirement for Earth-type life is *active plate tectonics*, which requires a *large planet* with abundant internal energy sources from radioactive decay to keep the interior mostly liquid or semiliquid. The required minimum size is somewhere between the size of Mars, which has no tectonic activity, and Earth, which does. Since Earth and Venus are actually the largest of the terrestrial planets, we would normally expect only one or two planets in a planetary system to be the correct size.

Finding Extrasolar Planets

To understand the difficulty of finding planets of other stars, consider first that an Earthlike planet would be about one astronomical unit from its parent star. A typical distance to a neighboring star might be a million astronomical units so that the angle between the star and its planet will be very small, but that is *not* the main problem because modern telescopes can actually resolve much smaller angles than that. The problem is that the reflected light from the planet will be *too faint to see* against the glare of its parent star.

The solution to the problem is to watch what we *can* see, the exact velocity toward or away from us of the parent star. (That is done using the Doppler shift of the star's spectral lines, which we will discuss later.) Because of Newton's Laws of Motion, any planets in orbit around the star will cause it to wobble in reaction and the amount of the wobble will tell us the mass of the planet that is causing it. Several hundred extrasolar planets have now been discovered in this way.

A drawback of the "wobble" method of finding extrasolar planets is that we usually find only the largest ones. Planets as small as Earth do not cause large enough wobbles to be seen easily.

Gliese 581c,d

Two planets of the star Gliese 581 have been discovered and are both of interest as possibly habitable planets. Gliese 581c was announced on April 24, 2007 and Gliese 581d on April 27, 2007.

Gliese is a red (type M3V) star, 20.4 light years away in the constellation Libra. The interest in Gliese 581c is that it is the smallest planet yet found to be anywhere near the habitable zone of its star. However it is still at least five times as massive as Earth and is quite close to its primary, taking only 12.3 days to go once around it. It is now thought that Gliese 581c is actually too close to its primary to be habitable. The intensity of light hitting the planet is about four times what Earth gets from its Sun, so it is probably like a large version of Venus.

Gliese 581d is actually a bit larger than 581c, more than seven times the mass of Earth, but is farther from the primary star and takes 83.6 days to go around it. It was, at first thought to be too far away from its primary star to support life but model calculations now appear to show that a greenhouse effect could easily increase its temperature enough to allow liquid water to exist on its surface. No direct observations of water have been made yet, however. For that, we would need the planet to pass in front of its star so that its atmosphere can produce an absorption spectrum for us.

The Star System Next Door

The closest stars to Earth form a trinary system just 4.4 light years away. It consists of Alpha Centauri A and Alpha Centauri B, in elliptical orbit around each other and distant companion, Proxima Centauri. Alpha Centauri A is a near twin of our Sun while Alpha Centuari B is an orange-yellowish star that is a bit dimmer than our Sun. These two orbit one another in elliptical fashion with a maximum distance of 35.6au and a minimum distance of 11.2au, completing an orbit in about 80 Earth years. Proxima Centauri is about 13,000au from the others and is actually the closest star to us.

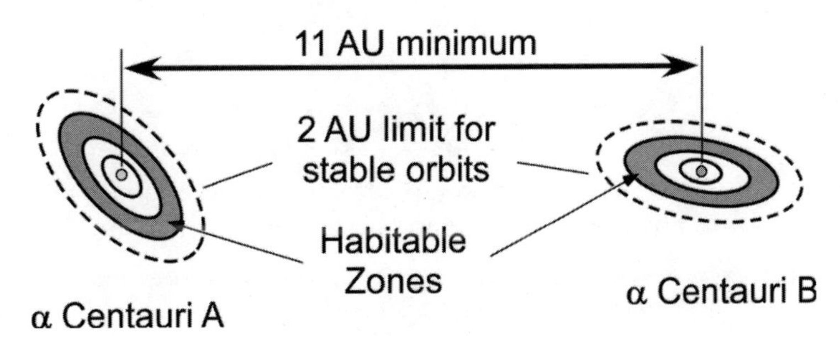

The main concern in a multiple star system is that the shifting gravitational fields of the orbiting stars may destabilize the orbits of planets. In this case orbits in the habitable zones of both Alpha Centauri A and Alpha Centauri B are quite stable and it is reasonable to expect there to be terrestrial planets in both zones. The closest approach distance of 11.2au would certainly destabilize any planets at the distance of Jupiter and the other gas giant planets. That is a concern since those Earth's water came from the rain of comets caused by those planets. However, there is a chance that, in the Alpha Centuari system, the two stars played that comet-throwing role for each other.

So far no planetary wobbles have been detected in these two stars. However, that is consistent with all of their planets being terrestrial with masses equal or less than that of Earth. In addition, (as of March 8, 2008) computer simulations of planet formation around Alpha Centauri B show that it may indeed have an Earth-mass planet orbiting in its habitable zone. Attempts to detect the wobbles that such planets would produce in Alpha Centauri A and B have begun, but will require several years of observation.

019.5 SETI: Search for ExtraTerrestrial Intelligence

The Interstellar Distance Problem

The nearest star system, Alpha Centauri, is 4.4 light years away. That means it takes light, or any other electromagnetic wave such as radio, 4.4 years to travel that distance. A radio conversation with someone there would have some long pauses but might still be worthwhile, particularly if they had a lot to say.

Sending probes to even the nearby stars is much more difficult. Current interplanetary space probes move at speeds up to about 50 miles per second, which is 0.02% of the speed of light. At that speed, travel to the nearest star would take 22,000 years. There are a variety of innovative ideas for getting to much higher speeds.

The Sun-diver would use current technology and take advantage of the fact that a rocket is much more efficient when it is already moving at high speed. In this approach, a gravitational sling-shot maneuver using one or more of the planets puts the probe on a path that brings it very close to the Sun. By firing its rockets just as it swings past the Sun, it can gain the maximum possible final speed. It might also jettison its rockets and fuel tanks and deploy solar sails to use the pressure of sunlight as it speeds away from the Sun. With this kind of scheme (and some wild optimism) one can imagine getting up to 500 miles per second or 0.2% of the speed of light. Now the trip time is down to only 2,200 years.

Fusion-powered Ion Rockets are far in the future, but one can work out their limitations and find that it will be very difficult to get them much above 10% of the speed of light. At that speed, the Alpha Centauri run will take 44 years each way.

Evidently we are not going to be able to get probes to even the closer star systems any time soon. That will make it very difficult to inspect the planets of those stars for primitive life-forms.

Help from the other end

Instead of looking for primitive life, just look for intelligent, technology-using life that is actually trying to be found. The obvious way to be found is to broadcast some sort of message.

Now we have three problems:

1. What kind of radiation do we use
2. What frequencies do we listen to?
3. How likely is it that there really is anyone out there sending us a message?

The Water Hole

So far, it appears that the most energy-efficient way to send a message is to use radio waves, which answers question 1. The frequencies are limited by noise sources and absorption in both interstellar space and in the Earth's atmosphere.

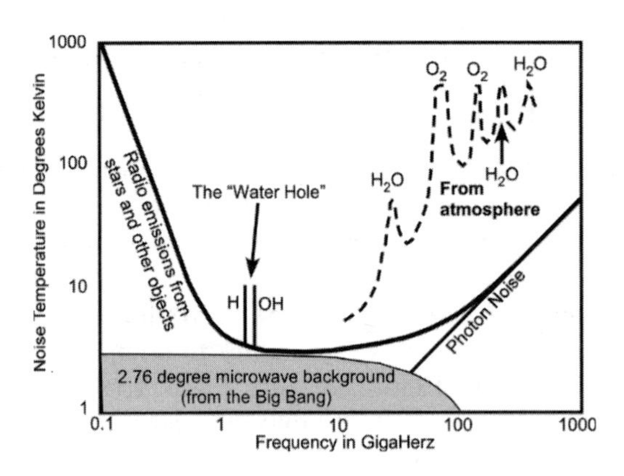

The vertical axis on these diagrams plots the amount of background noise as an effective temperature. As you can see, there is a nice minimum from 1 Gigahertz to about 10 Gigahertz. These are microwave frequencies, so the technology to listen to them is conveniently available.

That narrows down the answer to question 2, but we need to do better.

Within the quiet part of the radio spectrum there are two very prominent signals, one from the precession of interstellar Hydrogen and one from interstellar Hydroxyl ions. The region between these two absorption lines, from 1400MHz to 1727MHZ is called the **water-hole**. It has been argued that any water and carbon-based technological civilization would be expected to give this region of the spectrum special significance, so it is the frequency band that we should listen to for broadcast messages. That answers question 2.

The Drake Equation

The Drake Equation was originally conceived as a device for organizing discussions about the Search for Extraterrestrial Intelligence. It does not predict anything, but just puts everything that we know or suspect together. Frank Drake has referred to it as a way of "organizing our ignorance."

$$N = R^* \times f_p \times n_e \times f_\ell \times f_i \times f_c \times L$$

where

$N =$ Number of communicating civilizations in our galaxy

$R^* =$ Average rate of star formation in our galaxy

$f_p =$ fraction of stars that have planets

$n_e =$ average number of habitable planets in each planetary system

$f_\ell =$ fraction of habitable planets that actually develop life

$f_i =$ fraction of life-supporting planets that develop intelligent life

$f_c =$ fraction of intelligent species develop technological civilizations

$L =$ the average length of time that a technological civilization broadcasts signals

The values that summarized the 1961 meeting at which the Drake equation was conceived are as follows:

$$R^* = 10/\text{y}$$
$$f_p = 0.5$$
$$n_e = 2$$
$$f_\ell = 1$$
$$f_i = 0.01$$
$$f_c = 0.01$$
$$L = 10,000\,\text{y}$$

With these values, the Drake Equation gives

$$N = 10$$

communicating civilizations in our galaxy. That is a little depressing since our galaxy is 100,000 light years across and the chance of a close neighbor to communicate with is pretty small. Similarly, the broadcasters will have to be using **very** strong signals.

The value of R^* is based on direct observation and is pretty definite. The value of f_p, the fraction of stars with planets is still regarded as a good guess. The value of n_e, the number of habitable planets in a system, now looks *much too high.* The speed with which life arose on Earth suggests that the value of f_ℓ is about right. Whether life arose by abiogenesis or by panspermia, it appears to take hold as soon as the conditions are right for it. The values of f_i, f_c, and L are straight guesses based on no information at all.

The most vulnerable guess is probably the expected lifetime L of a communication-capable technological civilization. The amount of energy required for interstellar broadcasting implies the capability for efficient self-destruction by any number of different means. Our own civilization has had that status for about 100 years so far and it has not been easy.

It is not difficult to make very plausible changes in the values in the Drake equation that would make the value of N less than or equal to one. In that case, there is nobody out there to talk or listen to.

The Fermi Paradox

Fermi's question is "*Where is everybody?*" Our limited experience as a technological civilization suggests that such civilizations expand exponentially, doubling such parameters as population and energy use every 20 years and finding new resources to exploit whenever scarcity threatens. If such a civilization really lasts for even 10,000 years as the assumptions we put into the Drake Equation suggest, its impact on its environment should be *overwhelming and obvious.* If they do last that long, there seems no reason they should not last for millions of years, which would make them even more obvious — they should be visiting us in person by now.

An example of the sort of impact that we might expect to see is the *Dyson Star*. That is a star that has become so surrounded by a swarm of orbiting solar power plants that all we can see is the infrared radiation from their cooling fins. Such a thing would indeed be overwhelmingly obvious and unnatural.

Another example is related to our earlier suggestion that interstellar space is actually home to many non-stellar planetary mass objects that could easily become bases for settlement and exploration. Although interstellar travel between stars may always be forbiddingly difficult, travel between these waystation planets might be rather easy. In that case the conjectured 10,000 year old civilization would surely expand at perhaps 10% of the speed of light, occupying a sphere 1000 light years in diameter and swallowing all of the stars in that region into its power grid. There is no way that we could miss seeing such a thing.

In response to Fermi's asking where all of the extraterrestrial visitors are, Leo Szilard is said to have responded, "They are here, but we call them Hungarians."

Carrier Modulation: The Vanishing Earth Civilization

All of the various SETI attempts assume that aliens are sending a rather narrow band signal with information modulating a carrier wave. That particular technology is just now becoming outdated here on Earth as we move from analog TV broadcasts to digital broadcasts. Analog signals are *very easy* to spot because most of their energy goes into an obviously artificial carrier wave. Digital signals are useful precisely because they put much more of the signal energy into information content. That same feature causes them to look more like random noise to a receiver that does not know the correct decoding scheme. Analog Television signals have, for many years, been the strongest signal that we broadcast to the universe. As they are replaced by digital signals, the "Earth broadcast" is becoming less and less detectable by distant civilizations.

The lesson here is that an advanced civilizations may make less and less impact on its environment as it becomes more efficient. For example, it has been pointed out that a Dyson Star could have layers of power satellites with the outer ones using the emissions of the inner ones. Such a really efficient Dyson Star would be all but undetectable.

SETI Design

Since we do not know what frequency the aliens are using, we need to listen to as many as possible. That implies a multi-channel receiver that can listen on many frequencies at once.

The amount of information gathered by a multi-channel, multi-directional system is enormous, so a very high capacity computer is needed to process it all and pick out signals that might be of interest.

SETI Attempts

The OSU Big Ear Radio Telescope operated from 1963 until 1997. It consisted of a tiltable flat reflector sending radio waves to a fixed parabolic reflector that would focus them onto a microwave receiver. The Earth's rotation was used to scan in the east-west direction and the flat reflector was tilted to scan in the north-south direction. It was used for the longest running SETI project so far, from 1973 until 1997. In 1977 it received a signal that bore all the earmarks of being extraterrestrial. It is usually called the "Wow! Signal" because of the notation that a staff scientist made on the computer printout. The frequency and direction of the Wow! Signal have been under constant surveillance ever since then, but the signal has never repeated.

SERENDIP (Search for Extraterrestrial Radio Emissions from Nearby Developed Intelligent Populations) was started at UC Berkeley in 1979 and a successor program, SERENDIP II was begun in 1986 and was followed by several more upgraded programs. Each of these was a multichannel radio spectrometer that was attached to existing radio telescopes. The initial SERENDIP was a 100 channel analog radio spectrometer covering 100kHz of bandwidth. The most recent version, SERENDIP IV, consists of a 168 million channel spectrometer covering 100 MHz of bandwidth between 1.37GHz and 1.47GHz and has been installed and operating at the Arecibo radio telescope since 1999.The **SETI@home system** distributes chunks of the data obtained by SERENDIP to 5.2 million participating home computers. The program searches for

- spikes in the power spectrum corresponding to carrier waves.
- Gaussian rises and falls in power, corresponding to the telescope passing over a radio source as it turns.
- Triplets or three spikes in a row.
- Pulsing signals that might represent digital transmission.

So far, just one candidate signal was found on March 2003. That was Radio Source SHGb02+14a. The source was picked up three times at 1420MHz from a direction in which there are no stars for 1000 light years. Each time, its frequency started to drift in a fashion that is difficult to explain. It is generally thought to be of no significance.

019 Spot Check

Here are some questions to check your understanding of the material in module 019. Both the answers and where to find these questions at the website may found at the end of the Study Guide.

1 Which of the following statements best describes the presence of water on Mars?

a. Mars has liquid water flowing steadily on its surface at the present time.

b. Mars is completely without any form of water.

c. Mars has water frozen in its ice caps and may have liquid water below its surface.

d. Mars has never had liquid water on its surface.

2 The range of distances of a planet from its primary star that will permit the existence of liquid water on the planet's surface is called the

a. water hole.

b. triple point.

c. habitable zone.

d. inner system.

3 Here is the Drake Equation:

$$N = R^{*} \times f_p \times n_e \times f_\ell \times f_i \times f_c \times L$$

In this equation, L stands for the

a. number of extraterrestrial messages that we might expect to detect in a year.

b. expected lifetime of a communication-capable civilization.

c. average number of habitable planets in a planetary system.

d. number of communication-capable civilizations in our galaxy.

4 The main reason to suspect that Europa has a subsurface ocean of water is

a. patterns of cracks in the ice on its surface.

b. low fluxes of epithermal neutrons.

c. landmarks that are not rotating with the rest of the moon.

d. geysers of water shooting out through cracks in the moon.

5 A multiple star system close to the Earth that is known to have stable orbits in the habitable zones of each of its stars is

a. Albireo.

b. Epsilon Aurigae.

c. Sirius.

d. Procyon.

e. Alpha Centauri.

6 One difficulty with the microfossils that were found in the martian meteorite ALH84001 is that they

a. are the wrong shape to be living organisms.

b. could not exist on Mars at the present time.

c. are too small.

020: Stellar Parallax and Distance

020.1 Lenses and Mirrors

Optical Telescopes are basic tools of astronomy, so we need to know a little about how they work before we go any farther.

Focal Point of a Lens

A converging lens works by refracting light. Each part of the lens works like a prism and sends light toward the axis.

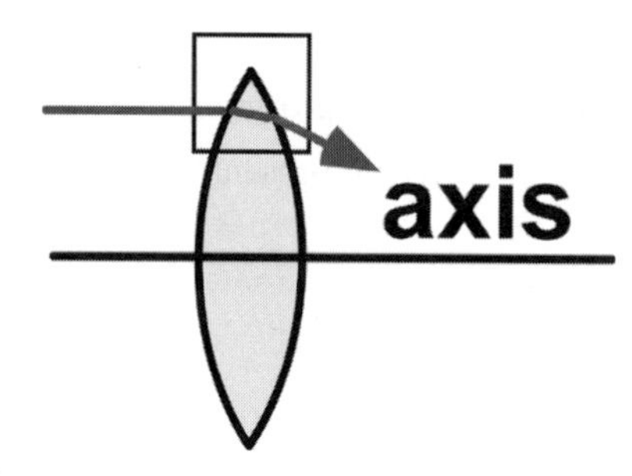

As a result, all rays parallel to the axis are sent through a single point, called the *focal point* of the lens.

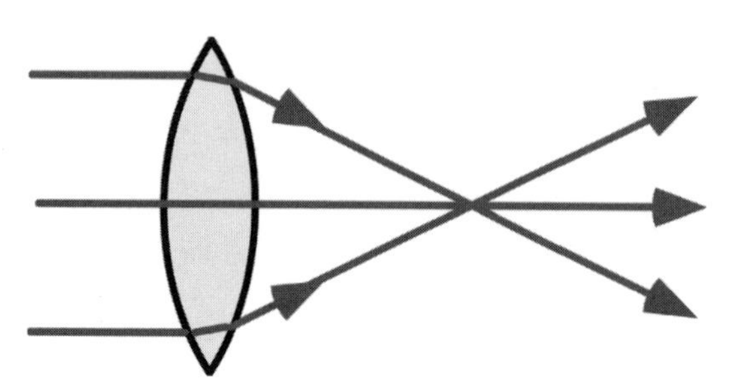

Because stars are very far away, the light rays from them are very nearly parallel. Light from a star that is exactly on the axis of the lens will be focused at the focal point.

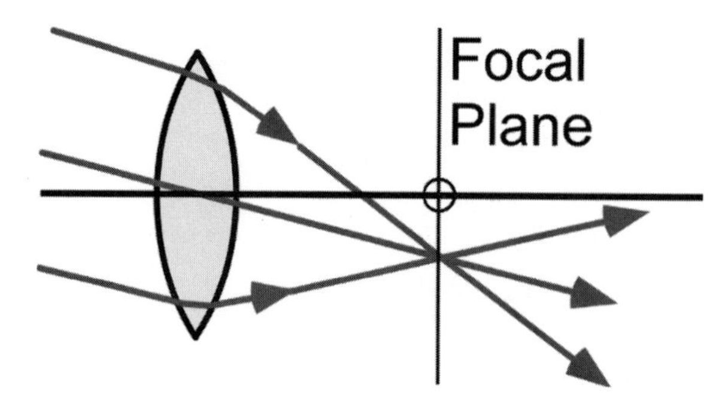

Light from a star that is off-axis will be focused somewhere in a plane that is perpendicular to the axis and crosses the axis at the focal point. That plane is called the *focal plane.*

Disadvantages of Lenses for Focussing Light

A major disadvantage of lenses is that the refraction of light depends on its wavelength.

Blue light is usually bent more than red light so that different colors focus at different points. This problem, called *Chromatic Aberration*, causes rainbow-like rings of color to appear around objects in images.

Lenses only work for wavelengths that can pass through glass. That limits them mostly to visible light. Infrared and ultraviolet light are absorbed by glass.

Because a converging lens depends on its shape to bend light, making it larger in diameter means making it thicker in the middle. The amount of glass that is needed increases very fast with diameter, thus limiting the practical sizes of lenses to about a meter in diameter. Beyond that size, they are simply too heavy to control easily and too difficult to cast.

Focal Point of a Mirror

Light reflects from a mirror in just the same way that a ball bounces from a surface. Both the incident ray and the reflected ray make equal angles with a line perpendicular to the mirror's surface and they both lie in a plane that includes the perpendicular.

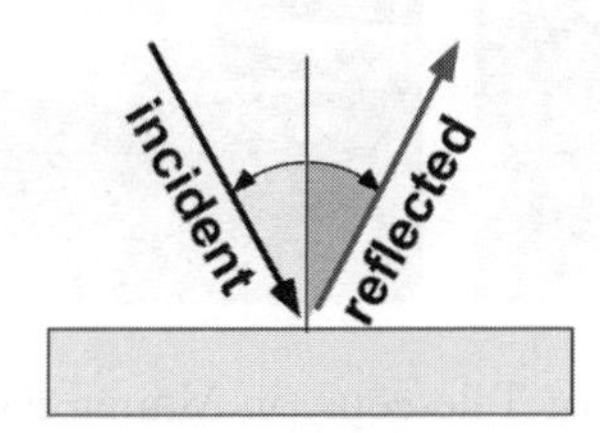

When its surface is curved, a mirror can reflect light in a way that depends on where the light hits the surface.

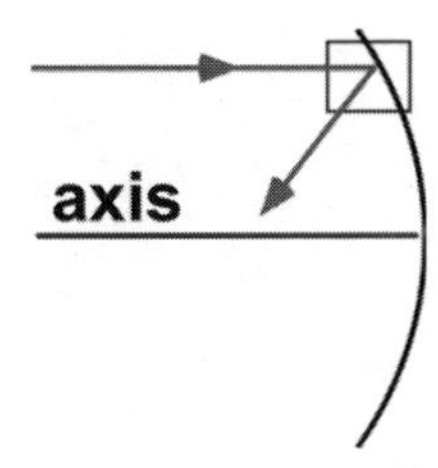

All rays parallel to the axis of the mirror are sent through a focal point much as in a lens, except that the focal point is on the same side of the mirror as the incident rays.

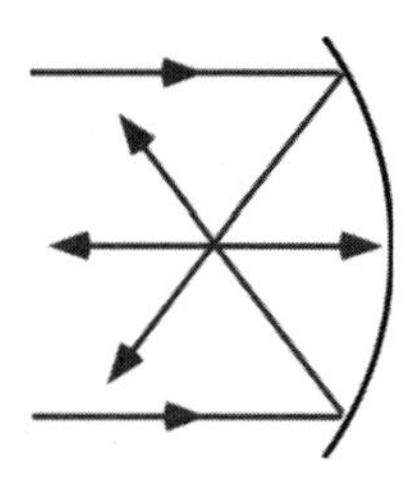

Advantages of a Mirror for Focussing Light

- All wavelengths reflect the same way. There is no chromatic aberration.
- Only one surface is needed, so the mirror does not have to become thicker when it becomes larger.

Mirrors can be up to five meters in diameter before their weight becomes a problem. Still larger mirrors, up to ten meters in diameter, are in use with computer-controlled mechanical actuators to maintain their shape.

Most large astronomical telescopes use a curved mirror to collect light.

These images are of the Subaru Telescope on Mauna Kea, HI.

020.2 Telescopes

Telescope Designs

An astronomical telescope is a simple device: A magnifier viewing the focal plane of a large concave mirror.

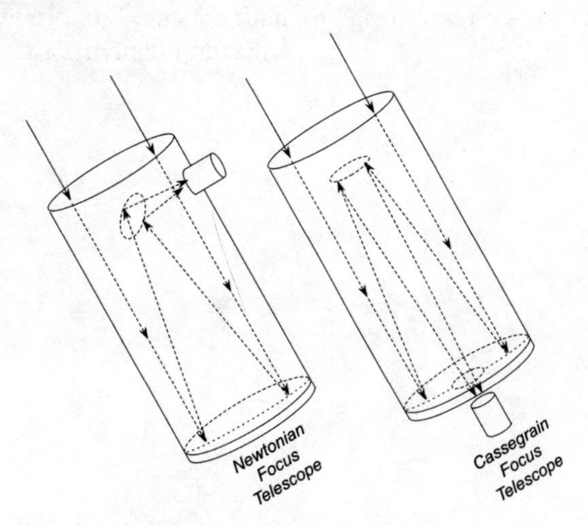

Smaller telescopes use either the Newtonian focus or the Cassegrain focus. Each has its own advantages and disadvantages:

Newtonian	Simple mirrors easy to build	Long, open tube vulnerable to dust and wind-pressure
Cassegrain	Hole in main mirror, convex secondary mirror	Short tube, usually sealed against dust

Larger telescopes use other variations on the focus:

- **Prime Focus** uses the natural focal plane and places the eyepiece in a cage inside the telescope tube.
- **Coudé Focus** adds a mirror to the Cassegrain design and diverts the light out the side to a location that can stay fixed as the telescope moves.

Very large telescopes usually use an open framework to locate the mirrors and rely on a building to keep out the wind and dust. The basic design has not changed much since the 1917 Hooker telescope. Its 100 inch diameter mirror was the state of the art at that time.

PLANCHE 31. — LE TÉLESCOPE HOOKER DE 2 m, 56 DE L'OBSERVATOIRE DU MONT WILSON.
PLATE 31. — THE 101-INCH HOOKER REFLECTOR, MOUNT WILSON OBSERVATORY.

The 200 inch (5 meter) diameter mirror of the Hale Telescope at Mt. Palomar was completed in 1948 and remains one of the world's most productive astronomical telescopes.

Large Mirrors

Mirrors larger than 5 meters in diameter begin to sag under their own weight. As the telescope moves to point in different directions, the sag changes and the mirror changes shape.

The cure for this problem can be seen in each of the four 8.2 meter diameter mirrors that make up the European Space Agency's VLT (Very Large Telescope). Each mirror is supported by a set of computer-controlled mechanical "actuators" that push and pull it to correct its shape as it moves.

Another approach is to divide the mirror into segments. The twin Keck Telescopes on Mauna Kea use mirrors that each consist of 36 hexagonal segments. Each segment is positioned by computer-controlled actuators so that they all function as a single mirror surface.

The Measures of a Telescope

A common mistake is to describe a telescope in terms of its *magnification power*. A 200X telescope would make images appear 200 times larger than the image that an unaided eye would see. The magnification power is determined by the eyepiece lens of the telescope and, for a given telescope, you can get *any* magnification that you want, just by changing the eyepiece. However, high magnification is not helpful if the image is too faint or too fuzzy to see.

Light gathering power is proportional to the area of the main mirror. The 200 inch Mt. Palomar Hale telescope has 4 times the light gathering power of the 100 inch Hooker telescope. A picture that requires a 15 minute exposure on the Hale would need a one-hour exposure on the Hooker.

Resolving Power is limited by the size of the mirror and the wavelength of light. When a telescope is pointed at a light source and turns a bit away from it, the light intensity that is sent to its focus does not drop off to zero immediately. The telescope must turn until its outer edge has moved through about one wavelength of the incoming light before the intensity drops. That one-wavelength turning angle is called the **diffraction limit** of the telescope's resolving power.

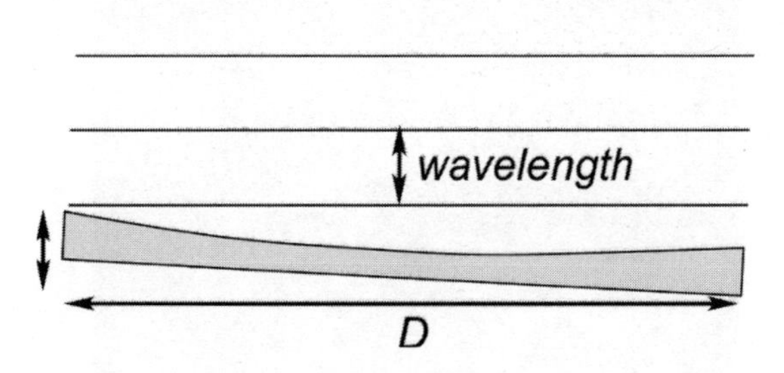

Overcome the diffraction limit by using several telescopes placed some distance apart. Combining the light from these different telescopes means that the one-wavelength turning angle becomes that of a mirror whose diameter is the distance between the telescopes. Such an arrangement is called an *interferometer.*

Atmospheric Turbulence

Earthbound telescopes such as the Hale at Mt. Palomar are not diffraction limited because they have a worse problem: Turbulence in the Earth's atmosphere. It makes stars 'twinkle' at night and causes the telescope image of a star to jump around. The image jumps too quickly for a photographic film to keep up with but not too quickly for computer controlled actuators to react. Using the actuators to distort the mirror to compensate for this jumping image problem is called *adaptive optics.*

Find a bright star and distort the mirror quickly enough to keep the image of that one "guide star" fixed in place. Other stars near the guide star will then be stabilized as well.

Create your own "guide star" by shining a powerful laser beam into the sky. The frequency is tuned to excite atoms in the high atmosphere. The glow of those excited atoms looks like a point of light and acts as an "artificial guide star."

Another way to avoid atmospheric turbulence is to put the telescope into orbit. The main mirror of the Hubble Space Telescope is 2.4 meters or 94 inches. It was limited by the size of the space shuttle cargo bay.

Although it is not impressively large, its diffraction limited resolution was, for may years, far better than that of any Earth-based Telescope. Because of adaptive optics, earthbound telescopes are now catching up to the HST. Future space telescopes will need much larger mirrors to compete. There is no limit to the possible sizes of orbiting mirrors because their support structures do not need to carry weight.

The Hubble Space Telescope was launched on April 24, 1990.

A correction for errors that were made in grinding the main mirror was made in 1993.

A final servicing mission is planned for August 2008.

The James Webb Space Telescope is scheduled to be launched in 2013. It will have a 6.5 meter diameter mirror and will function only in the infrared.

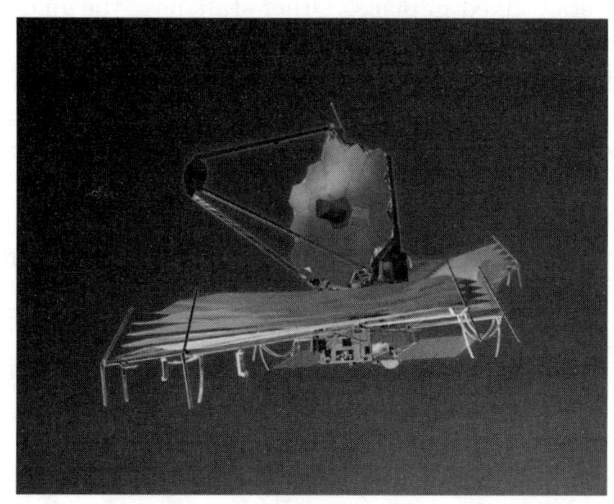

NASA

020.3 What Causes Parallax?

The Earth Moves

If we look at a nearby star at different times of the year, the Earth has moved around the Sun and our line of sight to that star changes.

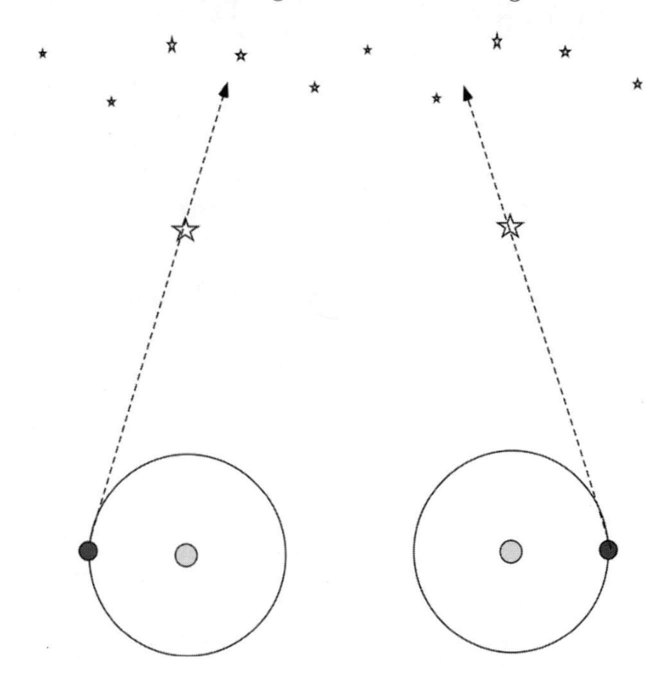

The nearby star appears to shift relative to the more distant stars that appear close to it in the sky.

Even the nearest stars are very far away, so the shift is too small to see without a good telescope.

How Large are Parallax Shifts?

The largest known parallax shift is that of Proxima Centauri: 1.52 seconds of arc.

Recall that a second of arc is 1/60 of a minute of arc which, in turn, is 1/60 of a degree.

The ancient Greeks understood that the motion of the Earth around the Sun should cause nearby stars to show parallax shifts.

Because they saw no such shifts, they concluded that the Earth cannot be moving.

Copernicus argued that the shifts are really there but the stars are so far away that they cannot be seen with naked eye measurements. He was right.

The best naked-eye angle measurements ever made were due to Tycho Brahe at his massive (for the time) observatory on the island of Hven. He could measure to one minute of arc or 60 seconds of arc.

It took until 1838 to find the first parallax. Friedrich Bessel found that 61 Cygni, a star that was known to be moving steadily due to its own proper motion, was also shifting back and forth by 0.6 of arc every year so its parallax angle was 0.3 second of arc.

020.4 Parallax Angle and Distance

The Triangle and the Parallax Angle

The parallax angle is defined to be half of the total annual shift in a star's postion.

That definition makes it one angle in a right triangle as shown below.

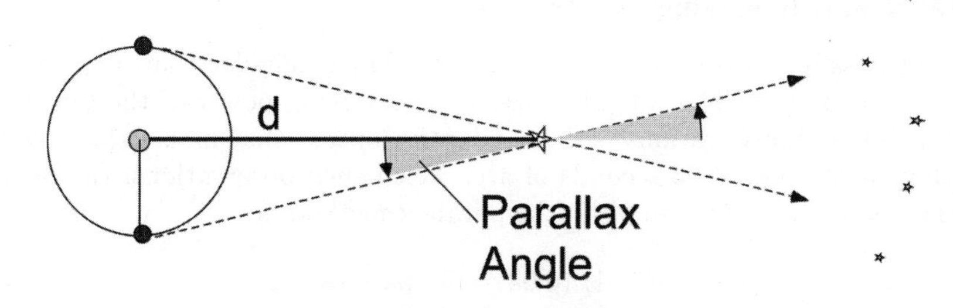

Notice that the distance is larger when the parallax angle is smaller.

The short leg of the triangle, half the distance between the two observation points is called the baseline. In the case shown, that distance is one astronomical unit: The radius of the Earth's orbit.

The Parallax Distance Formula

Because the parallax angle p is always extremely small, a simple formula connects it to the distance d between a star and our Sun:

$$d = 1/p$$

Distance Units in Astronomy

This formula gives the distance in *parsecs*. A parsec is defined to be the distance from our Sun to a star that shows a parallax angle of one second of arc when Earth's orbit is used as a baseline.

Another common unit for measuring stellar distances is the *light-year*, the distance light travels in a year. The relation between these units is:

$$1 \text{ parsec} = 3.3 \text{ light years}$$

Using the Distance Formula

The star Vega was found to show a parallax angle of 1/4 of an arc-second. Its distance from our Sun is then given by

$$d = 1/(1/4) = 4 \text{ parsecs}$$

or 4×3.3 light years = 13.2 light years.

The first parallax measurement, of 61 Cygni, was 0.3 seconds of arc. The distance from our Sun to 61 Cygni is then:

$$d = 1/\,(0.3) = 3.3 \text{ parsecs}$$

or 3.3×3.3 light years $= 11.1$ light years.

Distance and Resolving Power

It is not possible to observe a parallax shift that is smaller than the angular size of a star image. That angular size is the resolving power of the telescope. For Earth-based observations without adaptive optics, that makes the smallest parallax angle about 0.03 seconds of arc. With such observations, the largest distance that can be measured by the parallax method is

$$d = 1/\,(0.03) = 33 \text{ parsecs}$$

or 110 light years.

Space-based observations can resolve much smaller angles because they do not have atmospheric turbulence to contend with. The Hipparcos satellite measures parallaxes reliably to within 0.002 arc seconds. The corresponding distance is then

$$d = 1/\,(0.002) = 500 \text{ parsecs}$$

or 1650 light years.

020.5 Proper Motion

Stars are moving relative to our Sun. Over time, their position in the sky changes. This movement, in minutes of arc per year, is called proper motion.

Proper motion does not repeat annually the way that heliocentric stellar parallax does.

Proper motions are usually much larger than stellar parallax and were detected long before the first stellar parallax.

The Flying Star

The star 61 Cygnus has a proper motion of 5.22 arc seconds per year.

Because the motion does not repeat, it builds up year by year. In 100 years, 61 Cygnus moves across the sky by 522 arc seconds or about 8.7 mintes of arc.

The rapid proper motion of 61 Cygnus was well-known in 1838. It was called the "flying star". Friedrich Bessel reasoned that it must be passing close to our Sun and should show a large stellar parallax. He was right.

The small parallax shift of 0.6 seconds of arc each six months was superimposed on the much larger proper motion of 2.6 seconds of arc during that time.

020 Spot Check

Here are some questions to check your understanding of the material in module 020. Both the answers and where to find these questions at the website may found at the end of the Study Guide.

1 Barnard's star shows a proper motion of 10.36 arc seconds per year. In 100 years, its position in the sky changes by

a. 1036 seconds of arc.

b. 0 seconds of arc.

c. 10.36 seconds of arc.

d. 103.6 seconds of arc.

e. 518 seconds of arc.

2 You see a reflecting telescope with a short, stubby tube and the eyepiece at the back. This telescope uses the

a. Newtonian Focus.

b. Coudé Focus.

c. Cassegrain Focus.

d. Prime Focus.

3 The star Kruger 60 shows a heliocentric stellar parallax of almost exactly 0.25 seconds of arc. The distance from our Sun to Kruger 60 is

a. 2 parsecs.

b. 0.75 parsecs.

c. 4 parsecs.

d. 8 parsecs.

e. 0.25 parsecs.

4 A star is seen to move by 0.2 seconds of arc between February 1, 1999 and August 1, 1999 and then back to its starting point on February 1, 2000. What is the parallax angle for this star?

a. 0.5 seconds of arc.

b. 0.3 seconds of arc.

c. 0.1 seconds of arc.

d. 0.2 seconds of arc.

e. 0.4 seconds of arc.

5 A mirror that is supposed to bring light from a star directly overhead to a focus must be shaped like

a. a shallow trough with the open part facing up.

b. a flat surface.

c. a shallow bowl with the open part facing up.

d. an upside-down trough with the open part facing down.

e. an upside-down bowl with the open part facing down.

021: Using the Doppler Shift

021.1 Describing Waves

Definition

A wave is a pattern in the value of a quantity that is changing at every point of space.

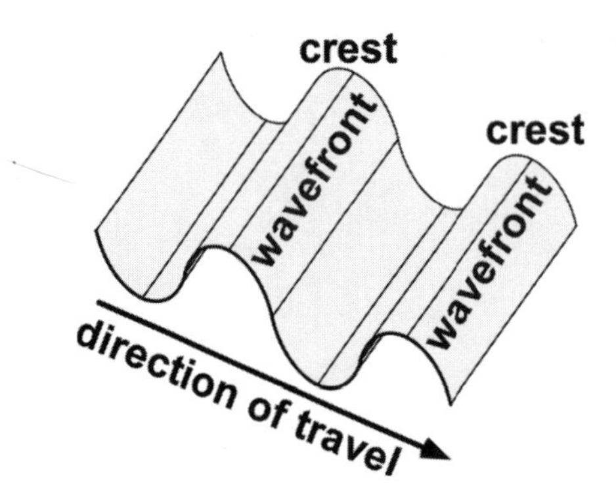

Wavelength

The distance from one wave crest to the next is called the wavelength.

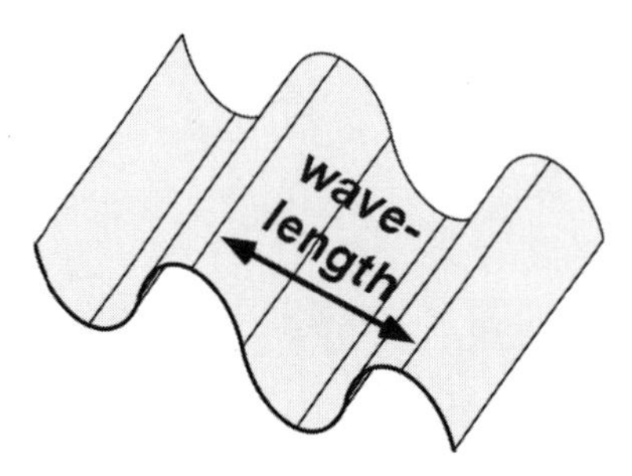

By convention, the Greek letter "Lambda" is used for wavelength.

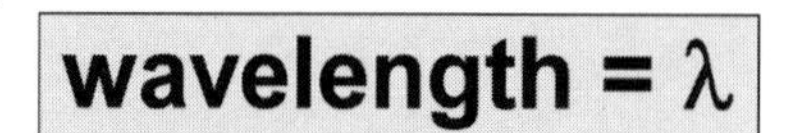

Sound

Sound consists of waves of air pressure. It can only travel where there is air.

The wavelength of the sound waves that make up middle-C is about four feet.

Light

Light consists of waves of electric field strength.

The electric field at a point of space just gives the force that a charged particle would experience if it were there. If there is no particle there, the field is still there. The electric field is a "what would happen if".

Light is a wave in a "what would happen if". It can travel where there is no matter at all.

The wavelength of red light is about 1.5 hundred-thousandths of an inch. That is the longest wavelength that our eyes can detect.

Units for Light Wavelengths

The customary unit for measuring light wavelength is the nanometer.

$$1 \text{ nm} = 10^{-9} \text{ meters}$$

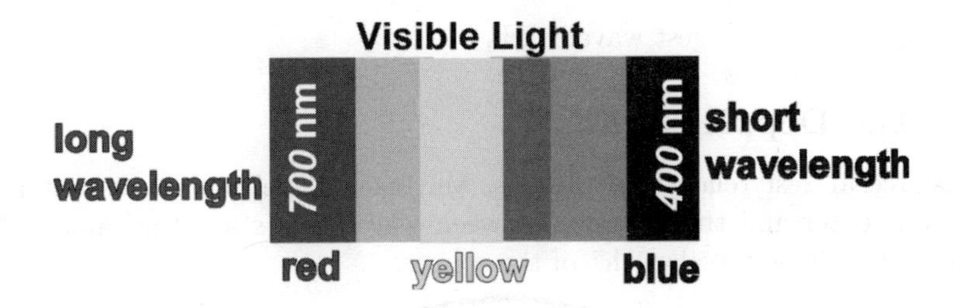

0.0.1 Speed of a Wave

The *velocity* or speed of a wave is defined to be:

Distance traveled by a wavefront divided by the time taken.

But the distance traveled by a wavefront is the same as the wavelength times the number of waves that pass.

$$\textit{Velocity} = \frac{\textit{Distance}}{\textit{Time}} = \frac{\textit{number of waves} \times \textit{wavelength}}{\textit{Time}}$$

$$= \frac{\textit{number of waves}}{\textit{Time}} \times \textit{wavelength}$$

$$\textit{Velocity} = \textit{frequency} \times \textit{wavelength}$$

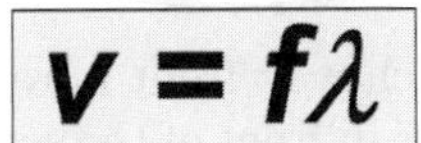

Speed of Light

In the absence of matter (in a vacuum in other words) light waves always travel at the same speed. Since that speed is constant and very special, we denote it by the letter c.

$$c = 3\times 10^8 \text{ meters per second}$$

Using the Speed Formula

Because the speed of light is always the same, the product of wavelength and frequency is always the same for light.

Violet light has the highest frequency of any light that our eyes can see. Red light has the lowest frequency.

Higher frequency means shorter wavelength, so violet light has the shortest wavelength.

Red light has the longest wavelength.

021.2 The Doppler Shift

For a source at rest relative to the observer, each wavefront moves out from a common center and the distance between wavefronts stays the same. This distance is the "true wavelength" of the source.

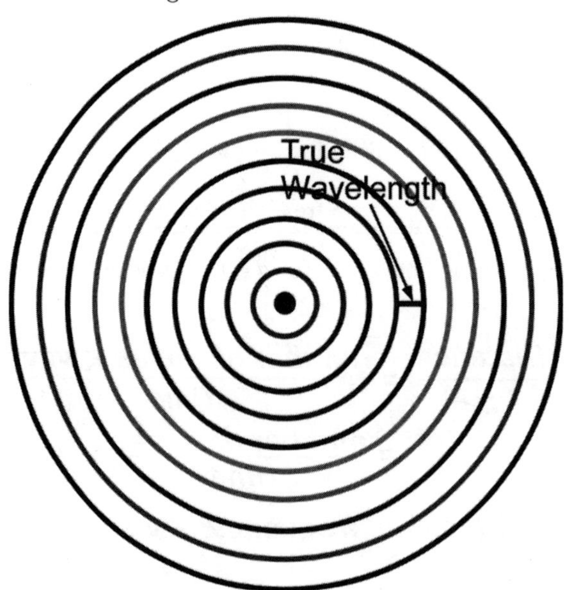

When a source is moving, each wavefront moves out from a new center and the wavefronts are closer together in front of the source and farther apart behind the source.

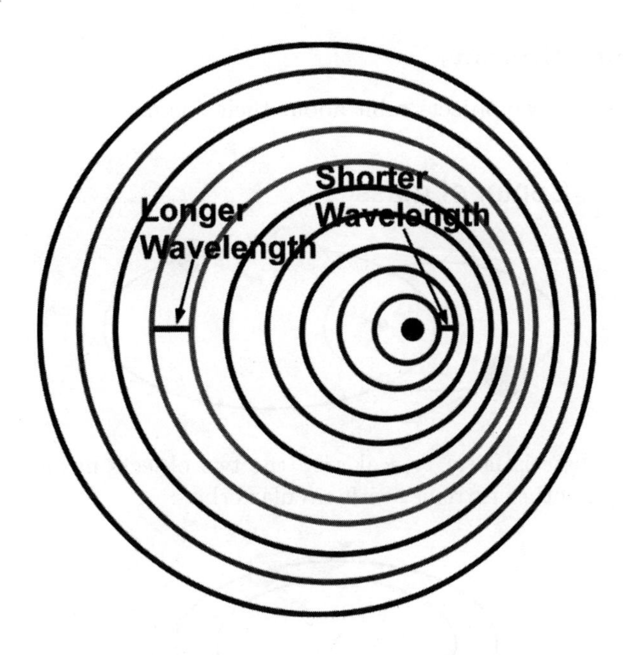

A spectrum that is shifted to shorter wavelengths is said to show a *blue* shift and indicates an approaching source.

A spectrum that is shifted to longer wavelengths is said to show a red shift and indicates a receding source.

From the amount of the wavelength shift, the actual velocity of the source toward or away from us can be calculated.

021.3 Binary Systems

In a binary system, two objects orbit about their common center of mass like this:

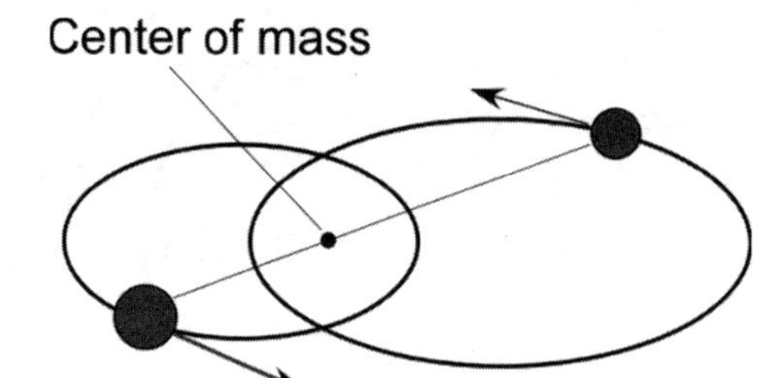

The case shown here is fairly typical with the two objects in highly elliptical orbits. Here is the other extreme, with circular orbits:

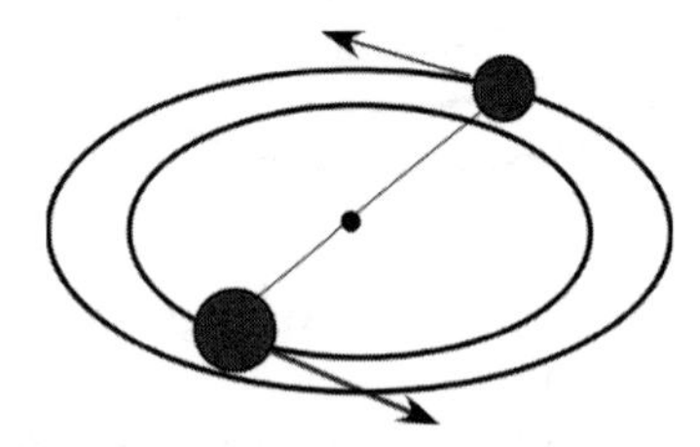

In this picture, the system is being viewed at an angle to the plane of its orbits so that the circular orbits look like ellipses. Notice, however, that they are arranged very differently from the eliptical orbits in the highly ellptical case. So long as we can see the details of both orbits, we can distinguish between circular orbits viewed at an angle and truly elliptical orbits.

Visual Binaries

When the members of a binary star system are well separated and the system is nearby we can sometimes resolve images of both stars. When we have a visual binary, we have a full picture of both orbits and can determine everything about it, by using Newton's Laws to find a model that fits what we see. We get both masses and the angle between our line of sight and the plane of the orbits.

Spectroscopic Binaries

In a binary star system, one star approaches us while the other one moves away.

The spectrum of such a system will show two different spectra, one for each star, with the spectrum of the approaching star blue-shifted and the spectrum of the receding star red-shifted.

Often, the stars of a binary pair are too close together to be resolved as separate images in a telescope. In such cases, the doubled spectrum is the only evidence that the system is a binary. Such a system is referred to as a *spectroscopic binary.*

The case that we described, where the spectra of both stars can be seen, is called a *double-line spectroscopic binary.* Often, however, one star is very faint, or in the case of an extrasolar planet, is not a star at all. In that case we see a single spectrum that shifts back and forth in a regular fashion. If we infer that the invisible partner is a star, the object is called a *single-line spectroscopic binary.*

When we try to fit a Newtonian model to a spectroscopic binary, not everything is determined. We have to *choose the angle* between our line of sight and the plane of the system. Unless we have some way of determining that angle, the best we can do is determine a *minimum possible mass* for each object in the system.

Ecllipsing Binaries

In a few cases, we just happen to be in the orbital plane of the binary star system that we are looking at. In those cases, we see one star pass in front of the other, causing a sudden dip in the total amount of light.

In an eclipsing spectroscopic binary we know that the angle between our line of sight and the orbital plane must be zero and can determine the masses of both objects.

By measuring the duration of the ecllipses we can determine the diameters of both stars.

By watching for changes in the spectrum during each ecllipse, we can learn something about the atmospheres of both objects.

021 Spot Check

Here are some questions to check your understanding of the material in module 021. Both the answers and where to find these questions at the website may found at the end of the Study Guide.

1 For an eclipsing spectroscopic binary star system, we can determine

a. the masses of both stars in the system.

b. the diameters of both stars in the system.

c. only the mass of the smaller star in the system.

d. only a minimum mass for each star.

e. the masses and diameters of both stars in the system.

2 Suppose that a sound wave has a wavelength of 12 meters and a frequency of 100Hz. What is the speed of sound?

a. 12 m/s

b. 8.34 m/s

c. 0.012 m/s

d. 1200 m/s

e. 100 m/s

3 The wavelength of the sound waves that correspond to middle-C is about 4 feet. If you are standing 8 feet away from a piano that is playing that note, then between you and the piano there will *usually* be

a. maximum pressure every two seconds.

b. maximum pressure every four seconds.

c. three regions of maximum pressure.

d. one region of maximum pressure.

e. two regions of maximum pressure.

4 In a particular binary star system, we are able to determine the masses of both stars in the system as well as the angle between our line of sight and the plane of the stars' orbits but cannot determine the diameters or atmospheric compositions of the two stars. This system is most likely

a. a spectroscopic binary system.

b. a visual binary system.

c. an eclipsing spectroscopic binary system.

5 The velocity of a wave is defined to be

a. the distance from one crest to the next.

b. the number of crests that pass divided by the time taken.

c. the number of crests that pass multiplied by the time taken.

d. the time taken for a crest to pass.

e. the distance traveled by a crest divided by the time taken.

6 Suppose that a star has a spectrum that includes red, blue, and violet lines spaced in the pattern of the lines from hydrogen but the violet lines are at 444 nm and 420 nm instead of the usual 434 nm and 410 nm. From this evidence, you can conclude that the star is

a. unusually cold.

b. moving away from us.

c. moving toward us.

d. rotating.

e. unusually hot.

022: Stellar Magnitudes and Distance

022.1 Luminosity, Brightness and Distance

Definitions

Luminosity is the total light output of the star: the energy emitted per unit time.

Apparent brightness is the energy per unit time per unit area that enters our telescope.

The Inverse Square Law

The light energy from a star spreads out equally in all directions. At a distance r from a star, the energy all passes through a sphere of radius r. The area of that sphere is $4\pi r^2$.

If the objective lens of our telescope has area A, then it does not intercept all of the energy from the star. Instead, it only intercepts the fraction

$$\frac{A}{4\pi r^2}$$

so we get the relation

$$\frac{\text{energy per unit time entering telescope}}{\text{luminosity}} = \frac{A}{4\pi r^2}$$

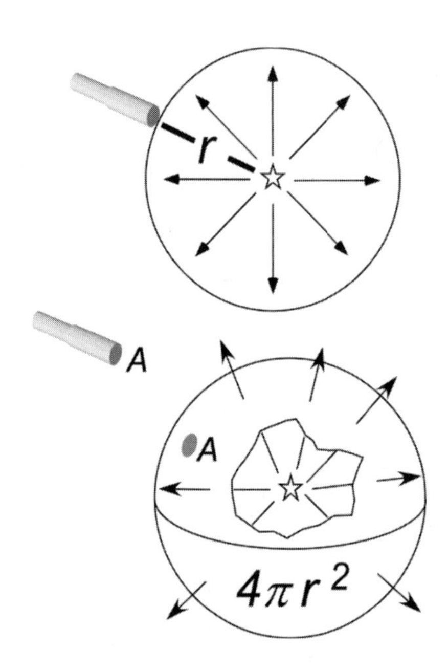

or

$$\text{energy per unit time entering telescope} = \frac{A}{4\pi r^2} \times \text{luminosity}$$

or

$$\frac{\text{energy per unit time entering telescope}}{A} = \frac{\text{luminosity}}{4\pi r^2}$$

or

$$\text{apparent brightness} = \frac{\text{luminosity}}{4\pi r^2}$$

This formula is called the *inverse square law for brightness.* It says that the apparent brightness falls off as the inverse square of the distance from a star.

Examples

If two identical stars are at different distances, they will have the same luminosity but different apparent brightnesses.

If one star is ten times as far away as the other, it will have only 1/100 of the apparent brightness of the closer star.

If one star is 100 times as far away as the other, it will have only 1/10000 of the apparent brightness of the closer star.

Notice the way the inverse square law works: If you *multiply the distance* by some number, you *divide the apparent brightness by the square* of that number.

Absolute Brightness

Instead of describing a star by its luminosity, astronomers find it convenient to use a related quantity, the *absolute brightness.*

The *absolute brightness* of a star is defined to be the apparent brightness that would be seen at a standard distance of 10 parsecs from the star.

(The choice of 10 parsecs as the standard distance is arbitrary and kind of annoying).

The advantage of using absolute brightness instead of luminosity is that we can forget about formulas and just use the way that the inverse square law works.

If a star is at a distance of 100 parsecs from us, we can compare its apparent brightness to the apparent brightness of a star at the standard distance. To go from the standard distance of 10pc to 100pc, you multiply the distance by 10. That means you divide its apparent brightness by 100. That gives

$$\text{apparent brightness at 100pc} = \frac{\text{apparent brightness at 10pc}}{100}$$

or

$$\text{apparent brightness at 100pc} = \frac{\text{absolute brightness}}{100}$$

Finding Distance the Hard Way

We can use the inverse square law to find the distance to a star. Shortly we will have an easy formula to plug into, but do it the hard way first.

A star is identified as being a certain type whose absolute brightness is known from observing nearby stars. It is found that the star's apparent brightness is 1/10000 of that absolute brightness. How far away is it?

We notice that in going from the standard distance of 10pc to the actual distance, the apparent brightness has been divided by 10,000 or $(100)^2$. Working the inverse square law backwards, we note that the distance must have been multiplied by 100. Multiply 10pc by 100 and get the distance to the star as 100×10pc or 1000 parsecs.

Here is a much easier example: A star is identified as a certain type whose absolute brightness is known. It is found that the star's apparent brightness is exactly equal to its absolute brightness. How far away is it? The answer, from the definition of absolute brightness, is that it must be at the standard distance of 10 parsecs.

The Key to Stellar Astronomy

These three things are connected:

Apparent brightness	How bright a star looks.
Absolute brightness	How bright a star really is.
Distance	How much the light spreads out while getting here.

A great deal of stellar astronomy consists of finding two of these things and calculating the third.

022.2 The Magnitude Scale

A Legacy from the Distant Past

The magnitude scale started out as *a subjective classification scheme* like this:

First	Wow! Look at that!
Second	Excellent!
Third	Uh huh.
Fourth	Pretty dim.
Fifth	Where is it?
Sixth	You're kidding me, right?

Because of the way that it started, the scale is *backwards.* Low magnitude numbers correspond to bright stars. High magnitude numbers correspond to dim stars.

The Modern Magnitude Scale

The scale has been modernized to relate it to measured brightness:

Add 5 to the magnitude —	Divide the brightness by 100.
Subtract 5 from the magnitude —	Multiply the brightness by 100.

As a starting point for the scale, the star Vega is arbitrarily assigned magnitude 0.0.

Adding just 1 to the magnitude corresponds to dividing by approximately 2.5. However, we will usually stay with multiples of 5 magnitudes to keep things simple.

This type of scale, with each increment multiplying the measured quantity instead of just adding to it, is called a *logarithmic scale.* It is often used to describe subjective judgements of quantities. For example, perceived sound intensity is measured in decibels where adding 10 decibels corresponds to multiplying the energy in the sound wave by 10.

Examples

A magnitude 1 star is 100 times as bright as a magnitude 6 star.

A magnitude 2 star is 100 times as bright as a magnitude 7 star.

The star Vega has magnitude 0, so it is 2.5 times as bright as a magnitude 1 star.

Compare the apparent brightness of a magnitude 11 star to the brightness of a magnitude 1 star. Note that going from magnitude 1 to magnitude 11 corresponds to adding 5 magnitudes *twice.* That then corresponds to dividing the brightness by 100 and then dividing it by 100 again. Thus, a magnitude 11 star has a brightness that is

$$\frac{1}{100} \times \frac{1}{100} = \frac{1}{10,000} = 10^{-4}$$

times the brightness of a magnitude 1 star.

Work this last problem the other way. Find the magnitude of a star that has 10^{-4} times the brightness of a magnitude 1 star. First note that

$$10^{-4} = \frac{1}{10,000} = \frac{1}{100} \times \frac{1}{100}$$

so that the brightness has been divided by 100 twice. That means 5 magnitudes have been added twice:

$$1 + 5 + 5 = 11.$$

Negative Magnitudes

The magnitude scale is extended to measure objects that are *brighter* than Vega by using negative numbers.

A magnitude -5 star would be 100 times as bright as a magnitude zero star such as Vega.

The planet Venus can get as bright as magnitude -4.4. That is bright enough to provoke a panic attack and is often the source of reports of unidentified flying objects.

Going increasingly negative corresponds to brighter objects. The Sun has magnitude -26.8 on this scale.

Naked Eye Limits

Under ideal conditions, magnitudes as high as 6 can be seen with a dark-adapted naked eye with perfect sight.

Under city lights, magnitudes above four cannot normally be seen with the naked eye. Some of the stars that make up the "Little Dipper" are fourth and fifth magnitude, so do not expect to see anything that looks like a little dipper from the city.

Seeing conditions can vary from one part of the sky to another. Thus, objects near the horizon are usually much harder to see than objects that are well above the horizon. Within the city, you may have trouble finding a magnitude 3 object that is near the horizon.

022.3 Apparent and Absolute Magnitudes

Definitions

The magnitude that describes the apparent brightness is called the *apparent magnitude.* It describes *how bright the star looks to us.*

The magnitude that describes the absolute brightness is called the *absolute magnitude.* It describes how bright the star would look from a distance of 10 parsecs and is a measure of *how bright the star really is.*

The difference

$$\text{apparent magnitude} - \text{absolute magnitude}$$

between the two kinds of magnitude is called the *distance modulus.*

Examples

A star with a distance modulus of zero would have to be at a distance of 10 parsecs from us.

Our own Sun has absolute magnitude 4.8. If we went 10 parsecs away from it and the looked back at it, we would find that it has apparent magnitude 4.8 so that it would be visible, but very faint.

The star Rigel, in the constellation Orion is already pretty impressive with apparent magnitude 0.18. However, its absolute magnitude is -6.69 so it would be stunning at the standard distance of ten parsecs. Evidently it must be a lot more than 10 parsecs away from us.

022.4 Finding the distance

The Hard Way

For each increase of the apparent magnitude by 5, divide the brightness by 100 and (from the inverse square law) multiply the distance by 10.

Suppose the apparent magnitude of a star is 22 while its absolute magnitude is 2. Find the distance to that star.

The reasoning goes like this:

At a distance of 10pc it would have apparent magnitude 2. To get to the actual apparent magnitude, add 5 four times.

$$22 = 2 + 5 + 5 + 5 + 5$$

That corresponds to dividing the brightness by 100 four times, or multiplying the distance by 10 four times. Thus, the distance is

$$10 \times 10 \times 10 \times 10 \times 10\text{pc} = 100,000\text{pc}.$$

The Easy Way

In terms of the distance modulus

$$DM = \text{apparent magnitude} - \text{absolute magnitude}$$

the procedure that we went through the hard way corresponds to the *distance formula*

$$d = 10^{DM/5} \times 10 \text{ parsecs}$$

This formula is really just a simple recipe for finding the distance to a star.

1. Subtract the absolute magnitude from the apparent magnitude.
2. Divide the result by 5.
3. Raise 10 to that power.
4. Multiply the result by 10 parsecs.

Try this formula for the example we did the hard way.

Suppose the apparent magnitude of a star is 22 while its absolute magnitude is 2. Find the distance to the star.

1. $22 - 2 = 20$.
2. $20/5 = 4$.
3. 10^4
4. $d = 10^4 \times 10$ parsecs $= 10^5$ parsecs $= 100,000$ parsecs.

Cases where it does not come out even

The distance formula is easy to use when $DM/5$ turns out to be an integer and that will be case for any exam problems that you might get. However, the formula still works for more general cases.

The distance modulus of the star Rigel is

$$DM = 0.18 - (-6.69) = 6.87$$

so that the distance to Rigel is

$$
\begin{aligned}
d &= 10^{6.87/5} \times 10 \text{ parsecs} \\
&= 10^{1.374} \times 10 \text{ parsecs}
\end{aligned}
$$

You can use a calculator to find

$$10^{1.374} = 23.659$$

so that the distance to Rigel is

$$d = 236.59 \text{ parsecs.}$$

The Distance to the Sun

Another example where the formula works even though the numbers look very odd is our own Sun. Its absolute magnitude is 4.8. Its apparent magnitude is –26.8. It distance modulus is then

$$DM = -26.8 - 4.8 = -31.6,$$

which looks a bit odd. The distance formula then gives the distance to the Sun in parsecs.

$$
\begin{aligned}
d &= 10^{-31.6/5} \times 10 \text{ parsecs} \\
&= 10^{-6.32} \times 10 \text{ parsecs} \\
&= 10^{-0.32} \times 10^{-6} \times 10 \text{ parsecs} \\
&= 10^{-0.32} \times 10^{-5} \text{ parsecs}
\end{aligned}
$$

Use a calculator to find

$$10^{-0.32} = 0.478\,63$$

or

$$d = 4.7863 \times 10^{-6} \text{ parsecs}$$

The distance to the Sun is the Astronomical Unit, so we get a connection between our two distance units:

$$1\text{au} = 4.7863 \times 10^{-6} \text{ parsecs}$$

or, solving the other way

$$1 \text{ parsec} = \frac{1}{4.7863 \times 10^{-6}} \text{au} = 208,930 \text{ au.}$$

Actually this connection between units can be derived strictly from geometry, with the exact result

$$1 \text{ parsec} = 206,265 \text{ au.}$$

so we are not too far off, considering that we gave the apparent magnitude (-26.8) to only three significant figures.

022.5 Preview of the Distance Ladder

Standard Candles

If you see a candle off in the distance, you know how bright it really is and you can see how bright it looks. You can calculate how far away it is. Of course, you have to be sure that it really is an ordinary candle and not, for example, a bonfire.

A recognizable light source whose absolute magnitude is somehow known is called a *standard candle.* Whenever we recognize one of these objects, we can measure its apparent magnitude and then calculate how far away it is.

Finding Standard Candles

The heliocentric parallax method of finding distance is very accurate for stars in our immediate neighborhood.

The Hipparcos satellite has measured accurate parallaxes, and thus distances out to about 1600 light years. There are approximately 15 million stars whose distances have been determined in this way.

For each of these neighboring stars we can measure the apparent magnitude and use the known distance to calculate its absolute magnitude. Thus we now have a sample of 15 million stars with *known absolute magnitudes.*

Before the Hipparcos satellite, the catalog of known absolute magnitude objects was much smaller since ground-based observations are limited to about 100 parsecs. However, that was still enough to show a close relationship between the absolute magnitude of a star and the sort of light the star produces — its *spectral type.* In fact, 90% of all stars show this close relation. As a result, 90% of all stars are usable as "standard candles."

The method of finding distances by using stars of known spectral type as standard candles goes by the very confusing name of *spectroscopic parallax.*

Moving Up the Distance Ladder

As an example of how the process works, suppose that we see a star whose spectral type is similar to that of our own Sun and has an apparent magnitude of 30. Calculate how far away the star is.

From the many thousands of stars with this spectral type that are in our neighborhood, we find that they all have absolute magnitudes close to 5. Thus the distance modulus for the unknown star is

$$DM = 30 - 5 = 25$$

and the distance is

$$d = 10^{25/5} \times 10 \text{ parsecs} = 10^5 \times 10 \text{ parsecs}$$

or

$$d = 10^6 \text{ parsecs.}$$

The star is a million parsecs away.

Limitations

The spectral types of stars can be recognized out to a distance of about 10 million parsecs, an enormous increase in range over the heliocentric parallax method.

To measure distance to objects that are even farther away, we look among the objects that are in our increased multi-million parsec neighborhood and seek standard candles that are bright enough to be identified at even greater distances. Objects such as Cepheid variable stars and supernova explosions take us up more rungs of the distance ladder.

At each step of the distance ladder, we make the assumption that distant objects are fundamentally similar to nearby ones. For example, we assume that an object with the same spectral type as our Sun is almost exactly like our Sun even though it is a million parsecs away.

That sort of assumption begins to break down when we get to objects that are so distant that their light was emitted billions of years ago when the universe was much younger than it is now. Errors from that source are called *evolutionary corrections*. To avoid (or at least detect) such errors, astronomers use a wide variety of different standard candle objects and seek a consistent interpretation of the results from all of them.

022 Spot Check

Here are some questions to check your understanding of the material in module 022. Both the answers and where to find these questions at the website may found at the end of the Study Guide.

1 A star whose apparent brightness is 10^{-4} times that of a first magnitude star would have magnitude

a. 21.

b. 11.

c. 1.

d. 16.

e. 6.

2 Suppose that the color and behavior of a star identify it as a type that we know has absolute magnitude –3. If the star's apparent magnitude is found to be 2, how far away is it?

a. 1000 parsecs.

b. 50 parsecs.

c. 100 parsecs.

d. 5 parsecs.

e. 10 parsecs.

3 Which of these answers describes the fundamental assumption that is behind all of the methods that astronomers refer to as the "distance ladder?"

a. Nearby objects show close relationships between absolute magnitude and spectral type.

b. Distant objects are similar to nearby objects.

c. It is possible to measure the apparent magnitudes of distant objects.

d. It is possible to calculate the absolute magnitudes of nearby objects.

4 Cruising far from the Sun, we notice that the Sun's apparent brightness has dimmed to 0.1 watts per square meter. We know that the apparent brightness at a distance of 1au is 1000 watts per square meter. How far from the Sun are we?

a. 1000au

b. 10au

c. 1au

d. 100au

5 A star with a distance modulus of zero is at a distance of

a. 10 parsecs.

b. 10,000 parsecs.

c. 1 parsec.

d. 100 parsecs.

e. 1000 parsecs.

023: Star Colors and Classes

023.1 Colors and Temperatures

Stars come in different colors. Here, in Orion, the star Betelgeuse (upper left) appears red to the naked eye while Rigel (lower right) appears blue-white.

Stars are extremely simple objects. Almost all of their properties are determined by their size and surface temperature.

The surface temperature of a star determines the relative intensities that it emits at different wavelength and thus, its color.

The hottest stars emit much of their energy in the ultraviolet and appear to be electric blue, like an arc-welder's torch. The star Sirius is a good example of a blue star.

The coolest stars emit most of their energy in the infrared and appear red, like a glowing electric stove element. The star Betelgeuse, in the constellation Orion is a good example of a red star.

Here are some typical surface temperatures and apparent colors of stars.

Surface Temperature (K)	Star	Color
30,000	Delta Orionis	Blue-UV
20,000	Rigel	Blue-UV
10,000	Sirius	Blue
7000	Canopus	Yellow
6000	Sun	Yellow
4000	Arcturus	Peach
3000	Betelgeuse	Red

For the hotter stars, the main differences in color are in the ultraviolet that our eyes cannot detect.

023.2 Spectral Types

How the sequence got so weird

Stars are classified according to the detailed appearance of their spectra. The important variable that determines that appearance is the surface temperature.

Before the physics of light emission was well understood, stars were classified according to their spectra in much the way that biologists would classify species of frogs.

The original classification scheme assigned letters of the alphabet according to the prominence of the emission lines of hydrogen in a star's spectrum.

Hundreds of thousands of stars were classified according to that scheme, so once it was realized that the important variable is actually not the hydrogen line intensity, but temperature, they did not re-name the types. They simply re-arranged them.

Arranged from the hottest to the coldest, the important spectral types are:

O B A F G K M

Several mnemonics for this sequence are available:

Traditional/Sexist	Oh Be A Fine Girl, Kiss Me!
Alternate/Sexist	Oh Be A Fine Guy, Kiss Me!
Weird Aussie	Oh Bring A Fully Grown Kangaroo, Mate!
Astronomers Rule	Only Bold Astronomers Forge Great Knowledgeable Minds.
Astro-Geek Chic	Optical Binary Affairs Fundamentally Generate Keplerian Marriages.

The spectral classes correspond to both overall color and surface temperature.

Spectral Class	Surface Temperature (K)	Star	Color
O	30,000	Delta Orionis	Blue-UV
B	20,000	Rigel	Blue-UV
A	10,000	Sirius	Blue
F	7000	Canopus	Yellow
G	6000	Sun	Yellow
K	4000	Arcturus	Peach
M	3000	Betelgeuse	Red

Notice that astronomers list the types from hot to cold which is backwards from what any sensible person would do. Recall that the magnitude scale is also backwards.

Examples

Which has a higher surface temperature, a type A star or a type B star?

Recall the sequence of types OBAFGKM. The sequence goes from hot to cold and B comes before A, so a type B is hotter than a type A.

Which has the highest surface temperature, on the list ABFGO?

Recall the sequence of types OBAFGKM. The sequence goes from hot to cold and O is first, so it is hotter than anything.

What would be the color of a type M star?

Recall the sequence OBAFGKM. It goes from blue, for the hottest stars, to red for the coolest. Since M is last, it must be red.

What would be the color of a type O star?

Recall the sequence OBAFGKM. It goes from blue, for the hottest stars, to red for the coolest. Since O is first, it must be blue.

023.3 Spectral Subclasses

Each spectral class is subdivided into ten numbered subclasses (0-9) according to surface temperature.

Like most things in astronomy, the order is backwards: The lowest number corresponds to the hottest star.

A type O0 star would be the hottest possible.

A type M9 would be the coolest possible star.

Examples

The star Mintaka (Delta Orionis at one end of Orion's belt) is a type O9. Mintaka is hot, but not the hottest possible.

The star Vega is a type A0 and is therefore hotter than Sirius, which is a type A1.

Arcturus, a K2, is hotter than Aldebaran, a K5.

Betelgeuse, a type M2, is hotter than Barnard's star, which is type M5.

Both our own Sun and the nearby star Alpha Centauri A are type G2.

023 Spot Check

Here are some questions to check your understanding of the material in module 023. Both the answers and where to find these questions at the website may found at the end of the Study Guide.

1 Which of the following spectral classes corresponds to the lowest surface temperature (on this list)?

a. G

b. A

c. B

d. K

e. F

2 Which of the following spectral types corresponds to the star with the highest surface temperature?

a. G5

b. G0

c. K0

d. K5

3 Which of the following colors indicates the coldest star?

a. peach.

b. red.

c. orange.

d. yellow.

e. blue.

024: The Hertzsprung-Russel Diagram

024.1 A dot for each star

For nearby stars, we can find distances from stellar parallax and calculate their absolute magnitudes.

For each star, draw a horizontal line across from its absolute magnitude and a vertical line up from its spectral type and place a point on the diagram.

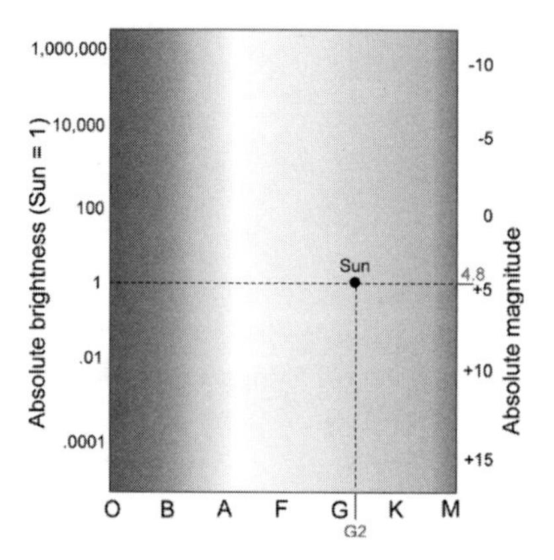

Each nearby star becomes a point on the HR diagram.

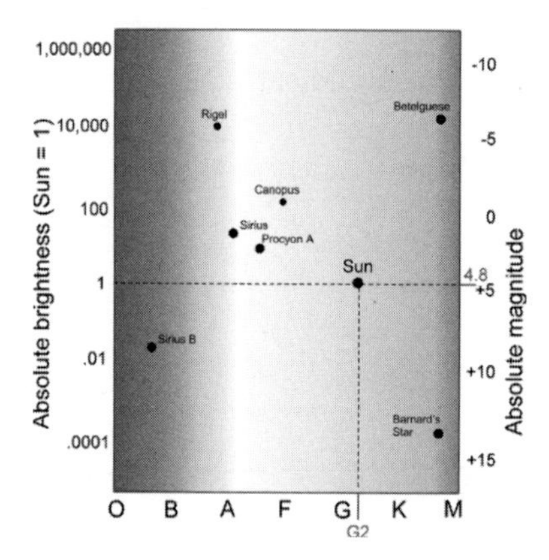

024.2 Interpreting the diagram

Astronomers are as backward as possible in this diagram. Temperature increases from right to left while magnitude increases from top to bottom.

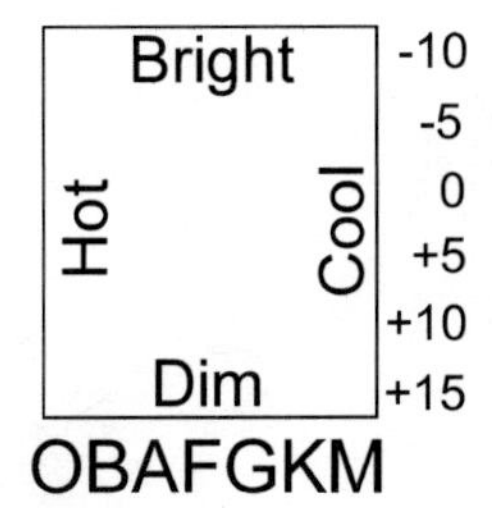

Since the magnitude scale is backward to begin with, brightness ends up increasing from bottom to top.

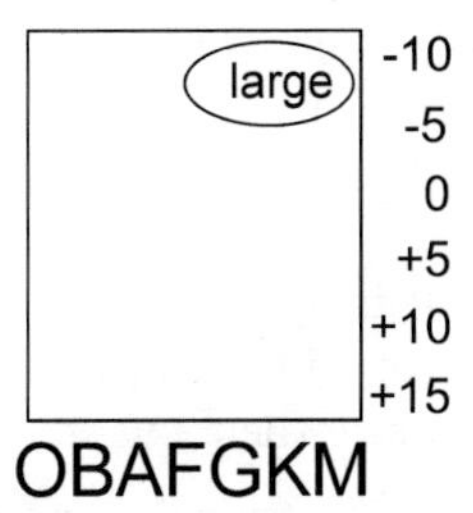

Objects in the upper right of the HR diagram are bright, but have low surface temperatures.

The only way that this is possible is if their surface areas are very large. Thus, these are "giant objects". Because they are red in color, they are called "Red Giants." Betelguese (the upper-left star in the constellation Orion) is an example of a red-giant star.

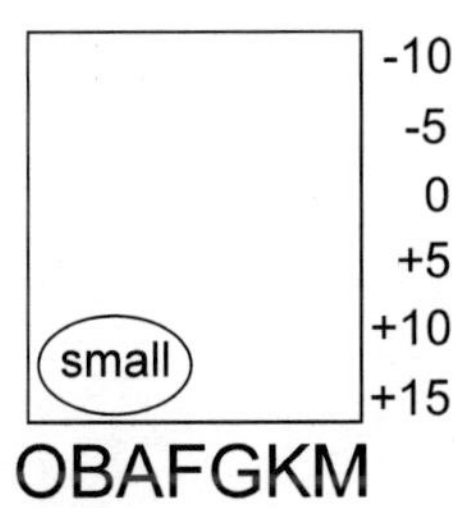

Objects in the lower left of the HR diagram are very hot objects which are also very dim. That situation can only happen if they have very small surface area. Thus, these are "dwarfs." The white dwarf companion Sirius B of the bright star Sirius is an example. Sirius is often called the "Dog Star" and its companion, accordingly, is called "The Pup."

024.3 The Main Sequence

When a large number of stars are located on the diagram, it is found that most of them lie close to an S-shaped curve that weaves from the lower right to the upper left.

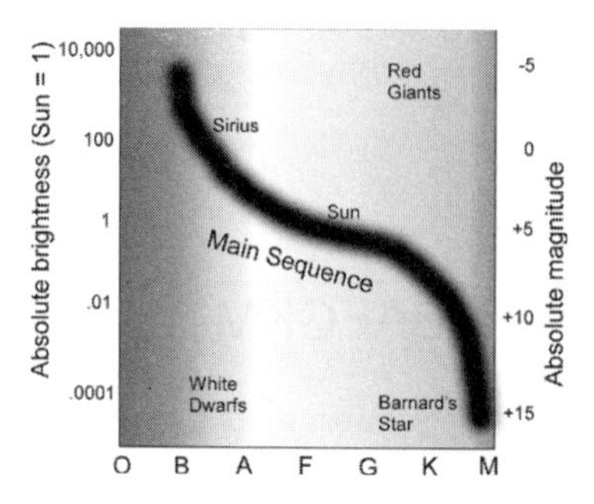

These stars make up the *main sequence.* Of the (roughly one million) stars that are close enough to the Sun for us to get stellar parallax distances, 90% are on the main sequence, nine percent are white dwarfs, and one percent are red giants.

By observing the doppler shift in spectroscopic binary stars, we can measure how fast the stars are moving around each other and can apply Newton's Law of Universal Gravitation to calculate their masses.

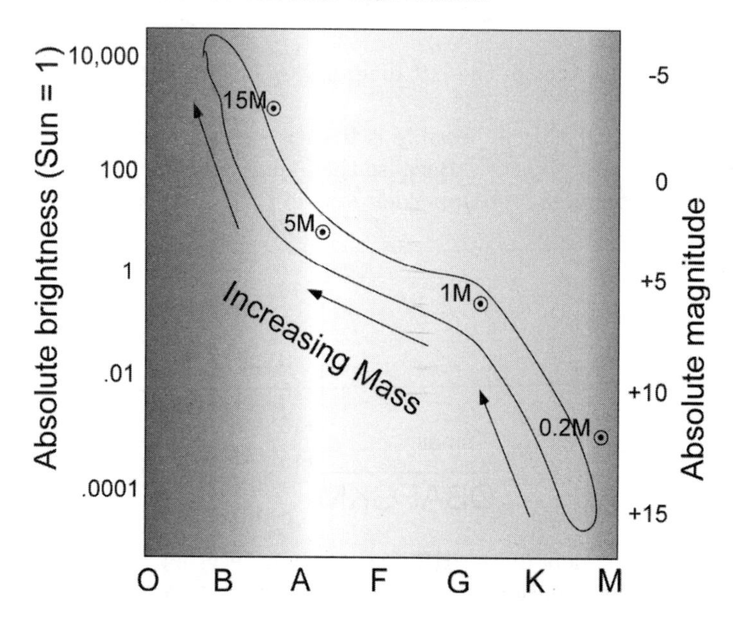

We find that the mass of a star is the only thing that determines its position along the main sequence. Stars with more mass than our Sun are brighter and bluer. Stars with less mass are redder and dimmer.

024.4 Luminosity Class

A star of a given spectral type could be a main sequence star similar to our Sun or it might be a red giant or a white dwarf.

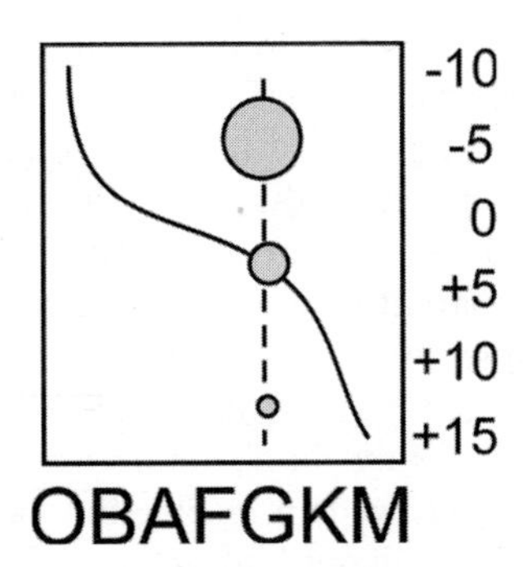

The gas in the photospheres of these different types of stars differs enormously in pressure. Higher pressure broadens the spectral lines, so we can determine it from the star's spectrum.

The pressure broadening of spectral lines determines a star's *luminosity class*, which, together with its spectral type, fixes its location in the HR diagram.

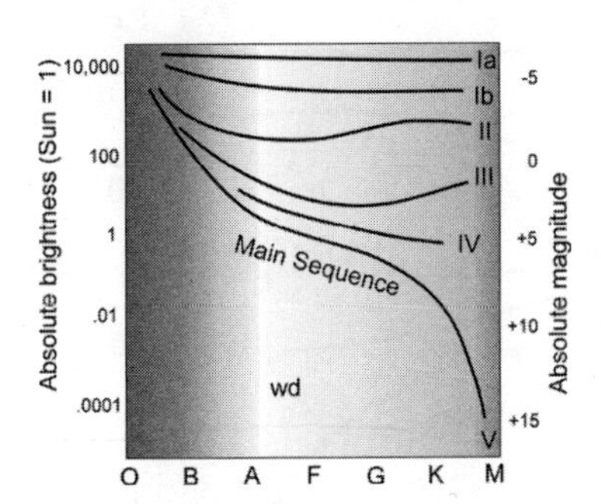

Class	Name
Ia	bright supergiant
Ib	supergiant
II	bright giant
III	giant
IV	subgiant
V	main sequence

The spectral class is usually combined with the luminosity class to give all of the information that can be determined from a star's spectrum.

The full spectral type of Rigel is B8Ia, a class B8 bright supergiant star: a blue supergiant.

The full spectral type of our Sun is G2V, a main sequence type G2 star: a yellow dwarf.

The full spectral type of Arcturus is K2III, which makes it a type K2 giant: a red giant.

White dwarf spectra are very different from the others because they have solid surfaces. They have their own luminosity class, usually denoted 'wd'.

A historical note: Henry Norris Russell came up with this sort of diagram in about 1910 after discovering one of the first known white dwarf stars. He wanted a way to express just how unusual it really was.

024.5 Spectroscopic Parallax

The Distance Ladder Again

The information about where the different luminosity classes are on the Hertzsprung-Russel diagram comes from observing the 15 million neighboring stars that we have good heliocentric parallax distances for.

Now that information gives us a way to measure *much larger* distances.

Given the full spectral type of a star, one can locate it on a Hertzsprung-Russell diagram and read off its absolute magnitude.

A measurement of the apparent magnitude of the star then gives the distance modulus and thus the distance to the star.

Direct stellar parallax is good out to about 1600 light years.

Spectroscopic parallax (a really dumb name for it, by the way) works to the maximum distance at which we can resolve individual stars and take their spectra. The limit is about 33 million light years.

Example

Suppose that a spectral type B6III star is found to have apparent magnitude 20. How far away is it?

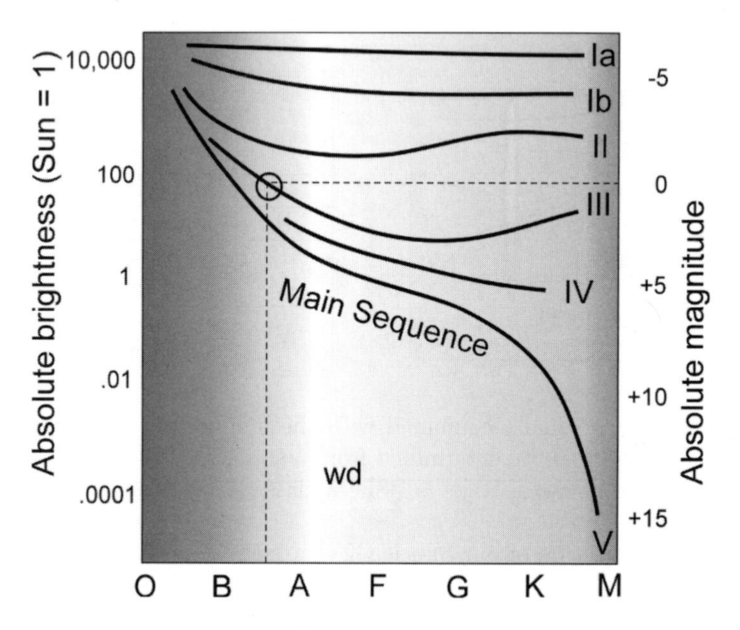

Draw a vertical line from type B6 up to the curve for luminosity class III and find that star has absolute magnitude 0.

Its distance modulus is 20-0 or 20.

The distance to the star is then

$$d = 10^{20/5} \times 10 \text{ parsecs} = 100,000 \text{ parsecs}.$$

024 Spot Check

Here are some questions to check your understanding of the material in module 024. Both the answers and where to find these questions at the website may found at the end of the Study Guide.

1 In the Hertzsprung-Russell Diagram shown, which point represents a star of type B with absolute magnitude +10?

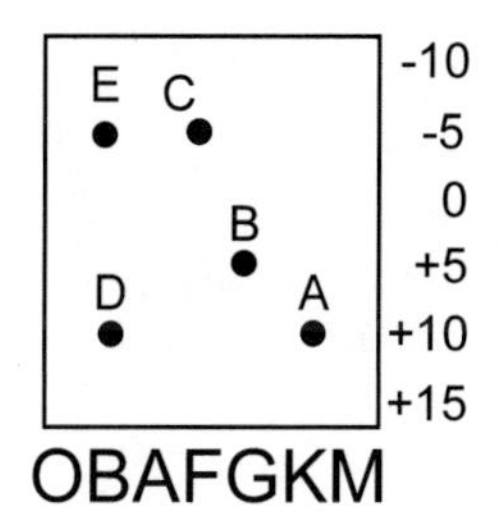

2 In a Hertzsprung-Russell diagram, the hottest stars are found

a. at the bottom.

b. on the left side.

c. at the top.

d. on the right side.

3 A main-sequence star with more mass than our sun will be

a. hotter and dimmer.

b. cooler and dimmer.

c. hotter and brighter.

d. cooler and brighter.

4 A star whose full spectral type is K2V is

a. a red giant star.

b. a bright blue supergiant star.

c. a red subgiant star.

d. a red main sequence star.

e. a red supergiant star.

5 Spectroscopic parallax uses

a. the doppler shift to find star velocities.

b. annual position shifts of stars to calculate distance.

c. timing variations in brightness to estimate mass.

d. stellar spectra to locate stars in the HR diagram.

025: The Births of Stars

025.1 The Building Blocks of Matter

The Cast of Characters

Types of Interaction

Particles interact with one another in several ways:

Force	Range	Effect
Electromagnetic	long	Like charges repel. Opposite charges attract.
Strong	short	Nucleons (protons or neutrons) stick together.
Weak	short	Particles change identity, generating neutrinos.

Interact through *all three forces*, strong, weak, and electromagnetic.

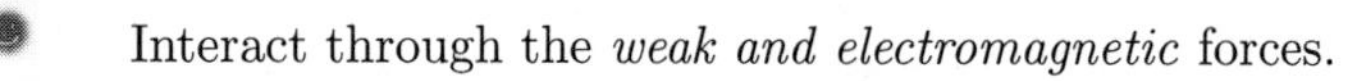

Interact through the *weak and electromagnetic* forces.

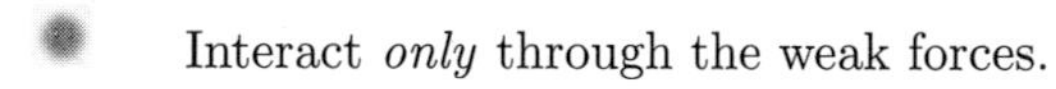

Interact *only* through the weak forces.

025.2 Mass and Energy

Conversion

The total amount of energy in a system never changes, it just changes from one form to another.

Mass is actually a form of energy. The formula for converting from the mass m (measured in kilograms) to energy E (measured in joules) is

$$E = mc^2$$

where c stands for the speed of light or 3×10^8 meters per second.

Particle-Antiparticle Annihilation

Positrons and electrons are exact opposites or antiparticles. When they meet, they annihilate each other.

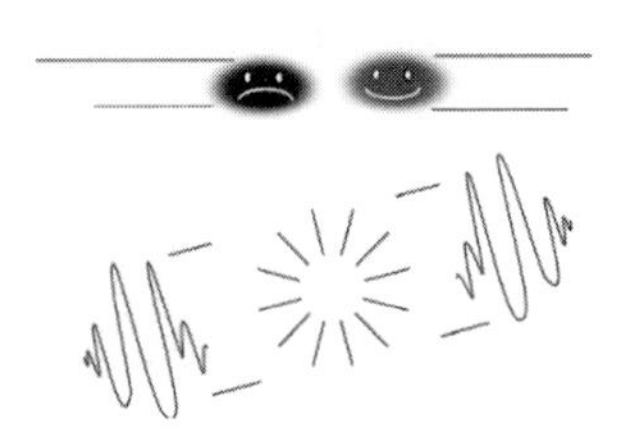

All of their mass vanishes and an equivalent amount of energy appears in the form of a pair of gamma ray photons.

Hydrogen Fusion

Our Sun, like most stars, consists mainly of hydrogen. This raw material undergoes a sequence of reactions, the net result of which is to form a helium-4 atom from four hydrogen atoms.

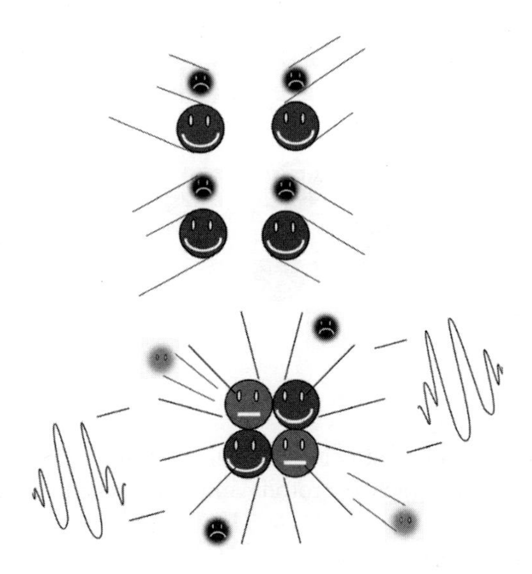

The process of building up heavier elements from lighter ones in this way is called *nuclear fusion.*

Compare the masses before and after.

Atom	Mass
H	1.008amu
4H	4.032amu
He	4.003amu

The difference, 0.029 amu or 0.75% of the original mass becomes energy.

If one kilogram of hydrogen is converted to Helium, 0.0075kg is converted into energy or

$$0.0075 \times \left(3 \times 10^{8}\right)^{2} = 0.0075 \times 9 \times 10^{16} = 6.75 \times 10^{14}\text{Joules}$$

In more familiar units, this amount of energy is 187,500,000 kilowatt hours. At a typical rate of ten cents per kilowatt hour, an electric power company would charge $18,750,000 for that much energy.

All main-sequence stars, such as our own Sun, are powered by the conversion of hydrogen into helium. The huge amount of energy that is released by this reaction explains how a star can shine for billions of years before exhausting its hydrogen fuel.

025.3 Ignition

Plasma

Temperature is the energy of motion of individual atoms or molecules. At the high temperatures in the core of a star, atoms move so energetically that their electrons cannot stay attached. The result is called a plasma.

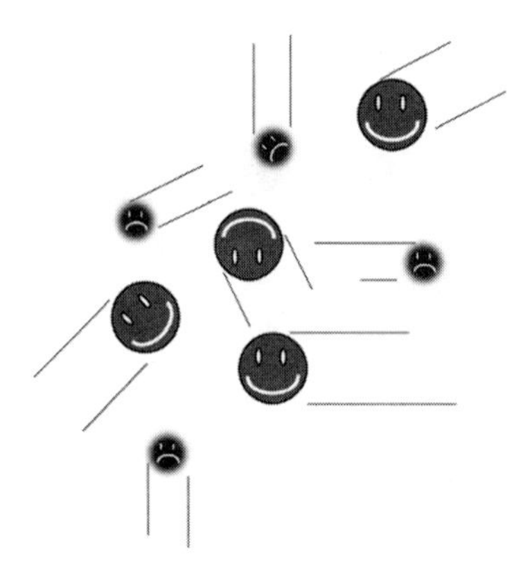

Most of the gas in a young star is hydrogen, so the plasma at the star's core is mostly a mixture of unattached protons and unattached electrons.

Nuclear Ignition Temperature

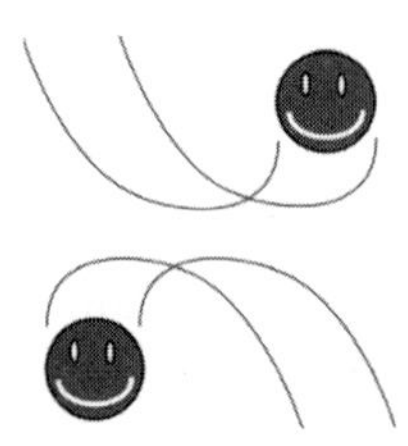

The positively charged protons repel each other. If the temperature is too low, the protons never get close to each other and nothing happens.

At *nuclear fusion ignition temperature*, the protons can get close enough to each other for the strong and weak nuclear forces to act and the protons can change into other particles.

A Star's Core is defined to be the region within which the temperature is above the ignition temperature for fusion: Roughly ten million degrees on the Kelvin temperature scale.

025.4 Evolution of a Protostar onto the Main Sequence

The Story in Words

Consider a cloud fragment similar to the solar nebula that formed our own Sun.

After about 100,000 years of collapse, the growing heat and pressure at the center of the cloud slow the collapse and the surface becomes hotter and brighter.

As the star collapses further, its surface area shrinks and it becomes less bright even though it is still getting hotter. This phase is called the Tau Tauri phase. It is unstable and gives rise to powerful protostellar winds of ejected gas.

After about ten million years, the central temperature rises enough to ignite nuclear fusion reactions. Over the following 30 million years, the star expands somewhat, then shrinks again and settles down for several billion years of life as a normal, main sequence star.

The Story in Pictures on the HR Diagram

Each stage in the evolution of a protostar has a brightness and a surface temperature, so it can be represented by a point on the HR diagram.

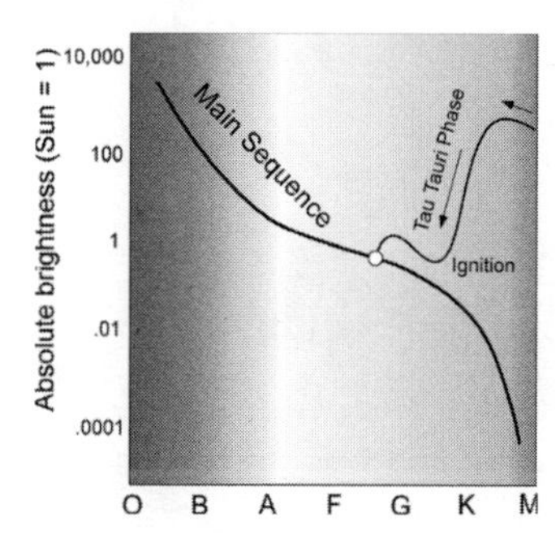

Here, the sequence of points for a one solar mass protostar is shown as it evolves toward the main sequence.

The size of the original collapsing cloud fragment determines the track of the protostar and where on the main sequence it stops.

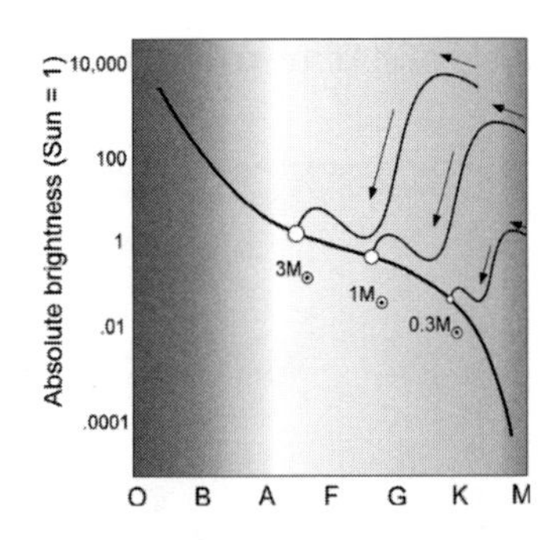

Once a star achieves stable nuclear burning, it sits at the same point on the main sequence for a very long time. In the case of a one-solar mass star, for billions of years.

A star at one point along the sequence never evolves into a star at another point. Instead, when the nuclear fuel runs out, the star moves off of the main sequence entirely.

025.5 Stars of Extreme Mass

The Lower Mass Limit

The lowest mass that can support nuclear burning is about 0.08 solar masses. Below that mass, protostars simply cool and contract to become brown dwarfs.

Failed stars or brown dwarfs are similar in size to Jupiter but are thought to be much denser and shine in the infrared with energy from their hot interiors. Below is a comparison between Jupiter, a typical brown dwarf star, and our Sun.

In the trapezium of the Great Nebula in Orion, infrared telescopes reveal an abundance of brown dwarfs.

The Upper Mass Limit

Stars that are more massive than our Sun become much hotter and brighter and evolve through all of their stages much more quickly. The most massive type O stars evolve about 50 times as fast as our Sun while the less massive M stars can take a billion years to reach nuclear ignition.

Above about 20 solar masses, stars become extremely unstable and only a few examples are known. At masses greater than about 50 solar masses, the light generated by the star exerts an outward force on the outer layers of the star that is greater than the force of the star's gravity. The star tears itself apart. Eta Carinae, at 100 solar masses, is tearing itself apart while we watch.

025 Spot Check

Here are some questions to check your understanding of the material in module 025. Both the answers and where to find these questions at the website may found at the end of the Study Guide.

1 Once a star has evolved onto the Main Sequence in the HR Diagram, it

a. drifts slowly toward lower mass and brightness.

b. stays at the same point until it runs out of fuel.

c. evolves up the sequence toward higher brightness.

d. moves both up and down the sequence.

2 An atom of ordinary hydrogen consists of an electron and a

a. deuteron.

b. proton.

c. neutron.

d. neutrino.

e. positron.

3 The mass of a carbon atom is 12.00amu while the mass of a helium-4 atom is 4.003amu. If three atoms of helium fuse to form carbon, how much mass is converted into energy?

a. 0.006amu

b. 0.009amu

c. 0.012amu

d. 0.002amu

e. 0.004amu

4 The average energy of motion of an atom or molecule in a gas is called its

a. speed.

b. temperature.

c. frequency.

d. entropy.

e. density.

5 Stars that are much less massive than our Sun

a. form more slowly but burn out faster.

b. form faster but burn slower.

c. form faster and burn out faster.

d. form more slowly and burn slower.

026: The Quiet Deaths of Ordinary Stars

026.1 Out of Fuel

The Fire Burns Outward

When the Hydrogen fuel begins to run out at the center of a star, only the product of nuclear burning, Helium, is left there.

The Helium core contracts and heats up.

The Hydrogen burning shell burns hotter and faster and closer to the surface of the star.

The star's photosphere expands by a factor of a hundred, becoming cooler and much larger.

The Story in Words

During the initial stage of expansion, the increase in surface area is almost exactly compensated for by the drop in surface temperature, so the star simply becomes redder but not brighter. During this early state, the star is said to be a subgiant and swells to about three times its size.

The star then swells to about 100 times its original size while its surface temperature drops very little. It is then called a red giant.

The Story in Pictures on the HR Diagram

The star evolves off the main sequence in two stages. The path shown is roughly the one that we expect our Sun to follow ten billion years from now.

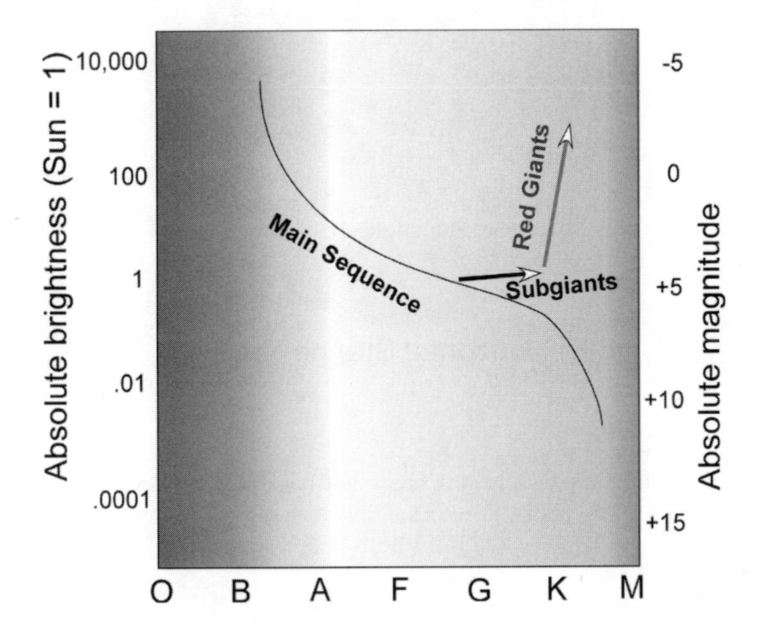

026.2 Second Chance

Central Heating Restored

During the red giant stage, the central helium core continues to contract and grow hotter until it reaches the ignition temperature for fusing helium to make carbon.

Because ignition occurs in a highly compressed core, there is an explosion, called the Helium Flash.

The new equilibrium, with energy coming from the center again, is more like a normal star, so the red giant contracts and its surface temperature rises.

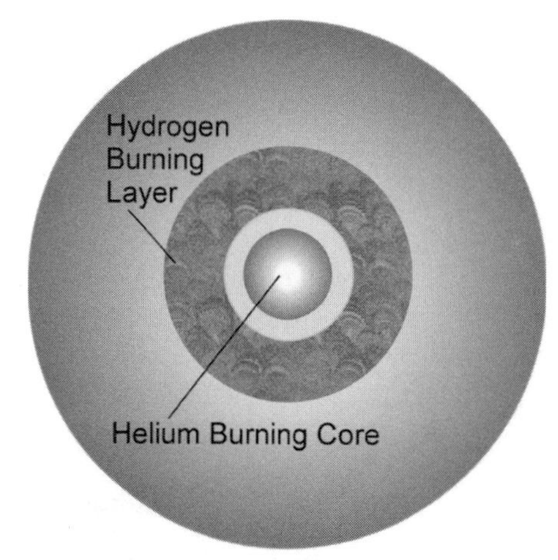

Horizontal Branch Star

The Story in Words

Once helium burning starts, the star has a hot central core as well as a shell of burning hydrogen. The result is intermediate between a red giant and a main sequence star.

The Story in Pictures on the HR Diagram

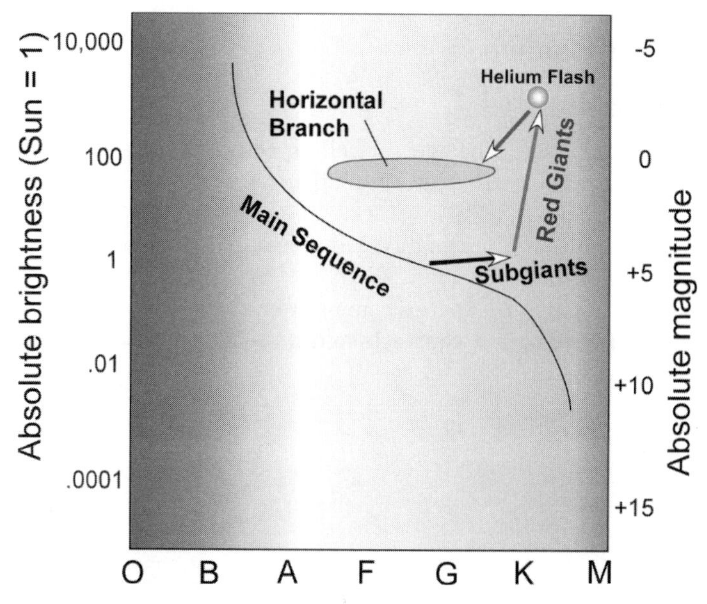

Stars in this intermediate state settle onto the Horizontal Branch in the HR diagram. They do not move along the branch but simply sit at one point on it.

026.3 Red Supergiant Stars

Soon, the helium begins to run out near the center of the star and a carbon core begins to build.

The helium-burning fire now burns outward and the star swells up one more time to become a red supergiant.

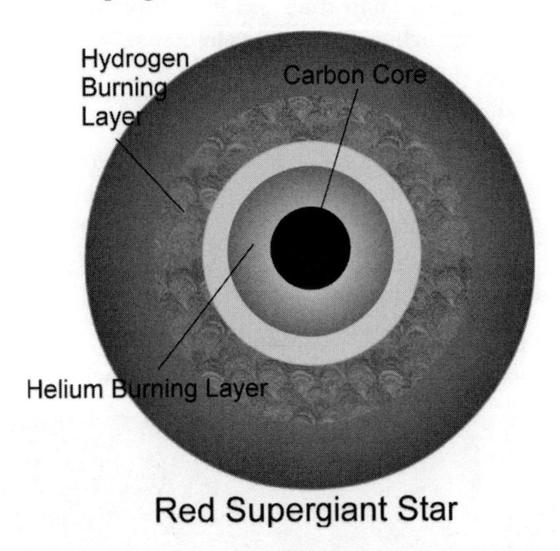

The second red giant phase is more violent than the first and the star becomes *much* larger. The star then leaves the horizontal branch and moves up to the top of the HR diagram.

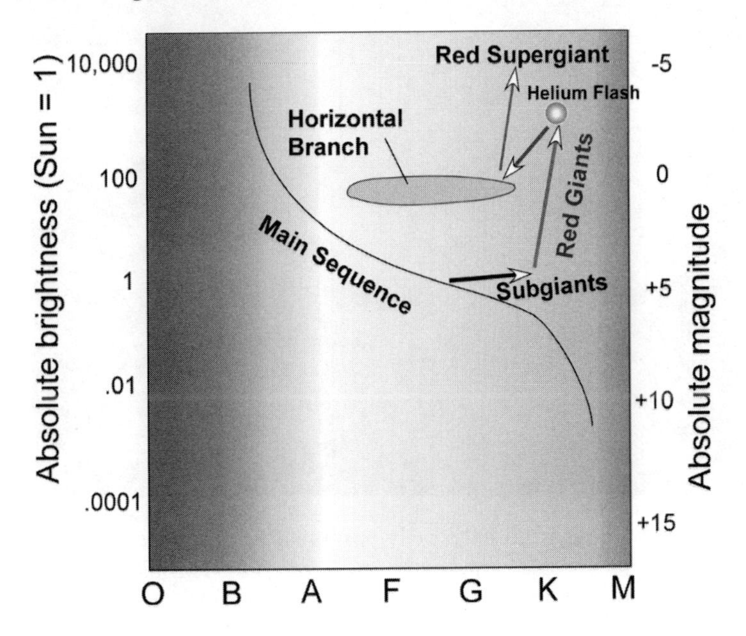

026.4 Game Over: Everybody Leaves

The Core Becomes Electron-degenerate

The carbon core shrinks to the point where *electrons touch each other.* That is called *electron-degenerate matter.*

It cannot shrink any more than that and its temperature never reaches the ignition point for burning carbon to heavier elements.

A Planetary Nebula Forms

The outer layers of the star eventually become unstable and blow away, leaving only the degenerate core. The hot core is now a white dwarf star and the outer layers have become a *planetary nebula.*

Examples of Planetary Nebulae

The Stingray Nebula is a binary star system in which one of the two companions has just lost its outer envelope and become a white dwarf star.

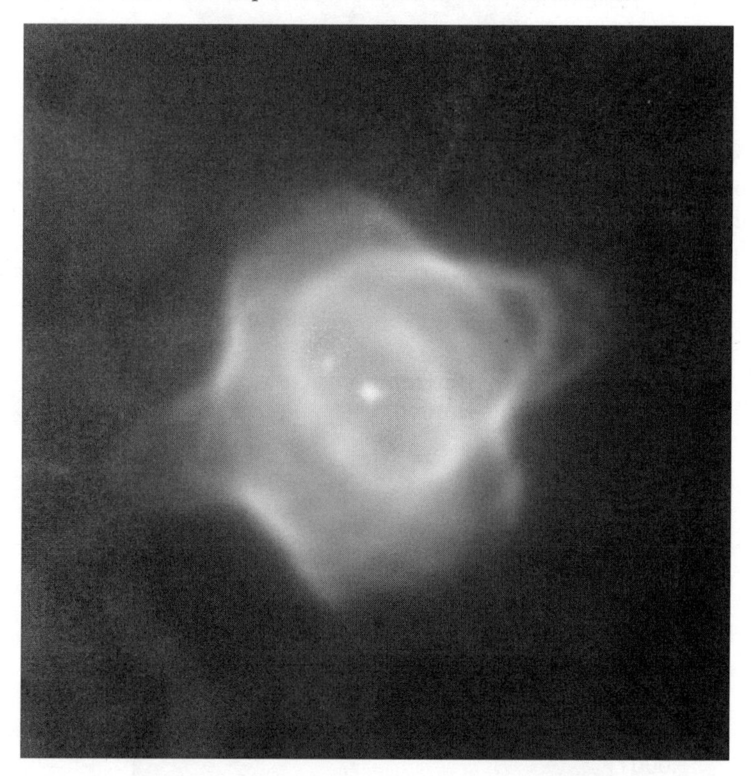

Planetary nebulae can be extremely complicated. Their shape depends on the way in which the star ejects its outer envelope. This Hubble Space Telescope image of the Hourglass Nebula (shown on our website) is a good example of complexity.

Here, (image shown on our website) a newly born white dwarf can be seen as the small blue dot at the center of the ring nebula.

Information from Planetary Nebulae

The gas in a planetary nebula is speeding away from the remaining white dwarf. The doppler shift in the spectrum of gas coming directly toward us tells us its speed so that we can figure out how long ago it started its expansion. Most planetary nebulae are just a few thousand years old. Eventually the expansion carries all of the gas away from the star and the planetary nebula fades away.

Planetary nebulae last only a relatively short time and each one represents the death of a star similar to our own Sun. The number of planetary nebulae that we see then tells us about the death rate for such stars.

White Dwarf Stars

The planetary nebula dissipates in just a few thousand years, leaving behind a white dwarf star about the size of our Earth but with a surface temperature in the millions of degrees and a mass similar to that of our Sun.

Eventually the white dwarf is alone, or perhaps orbiting a still-living companion star.

Sirius B, the "Pup" is a nearby white dwarf star orbiting the star Sirius. It's mass is 1.1 times the mass of our Sun and its radius is just 5500 kilometers. It resulted from the death of a star with about four times the mass of our Sun.

A white dwarf no longer generates energy and cools until it eventually becomes an invisible black dwarf star.

When it first forms, a "white dwarf" is actually blue and changes color through white and red as it cools. They are called "white dwarfs" after the first one that was noticed (which was also the original inspiration for Russell to introduce the HR diagram).

On the HR diagram, the dying star moves to the left as the hot interior is uncovered, then becomes dimmer as it contracts and eventually cools as it uses up its stored heat energy.

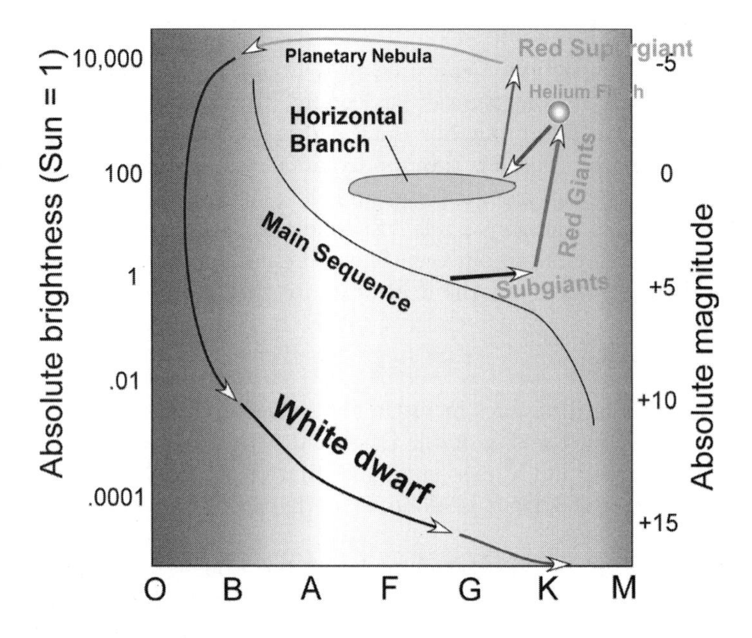

026.5 Return from the Dead: Novas

The Cause: A Felonius White Dwarf

Many white dwarf stars have normal companion stars. If the partner star is very close, then it can dump hydrogen gas onto the white dwarf.

The intense gravitational field of the dwarf causes the gas to fall onto the dwarf's surface with great speed, heating the surface until a nuclear fusion reaction is initiated. The result is a "new star in the sky" or nova.

The sudden flare of nuclear energy blows away the fuel and the process starts again. Some stars go nova over and over again.

Examples

Here is an artists impression of what is going on in the binary star system RS Ophiuchi. It is about 2000 light years from us and consists of a red giant and a white dwarf. About every 20 years, the white dwarf succeeds in getting enough hydrogen from its overblown companion to set off a thermonuclear reaction. The resulting flare makes this normally dim star visible to the naked eye.

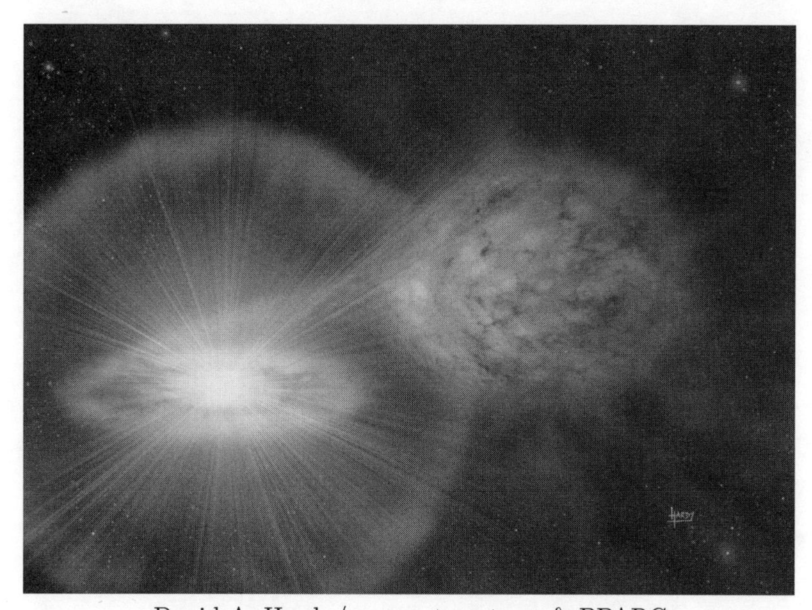

David A. Hardy/www.astroart.org & PPARC

In this Hubble Space Telescope image of the nova Cygni 1992, the material of the ring can be seen, from its spectra, to be nuclear reaction products.

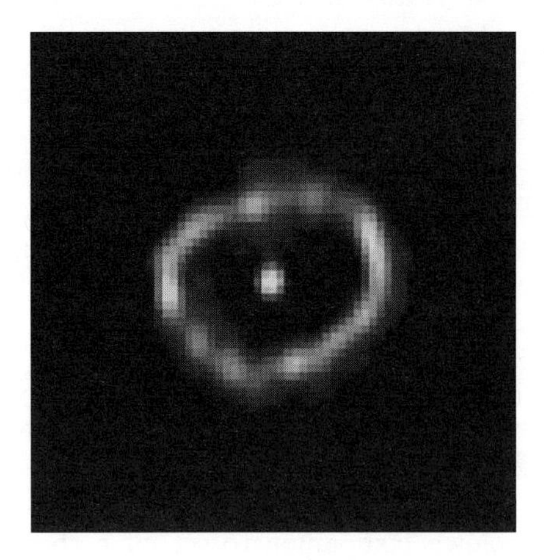

026 Spot Check

Here are some questions to check your understanding of the material in module 026. Both the answers and where to find these questions at the website may found at the end of the Study Guide.

1 A planetary nebula forms when

a. a supernova happens.

b. a new star is born.

c. a star becomes a white dwarf.

d. a nova happens.

e. a star begins to burn helium.

2 The red supergiant phase of a star is caused by

a. the collapse of its core.

b. the exhaustion of helium at its core.

c. the ignition of helium at its core.

d. the ignition of hydrogen at its core.

e. the exhaustion of hydrogen at its core.

3 The red subgiant stage of a star is best described by

a. increasing temperature and decreasing brightness.

b. constant temperature and brightness.

c. dropping temperature and increasing brightness.

d. dropping temperature and constant brightness.

e. increasing temperature and increasing brightness.

4 On a HR diagram, a visible white dwarf star is in the

a. upper right corner.

b. lower right corner.

c. main sequence.

d. upper left corner.

e. lower left corner.

5 Stars on the horizontal branch of the HR diagram are burning

a. helium in a shell around a carbon core.

b. helium at their centers.

c. hydrogen around a helium core.

d. hydrogen at their centers.

027: Supernova Explosions

027.1 Heavy Element Formation in Massive Stars

Building Up To Iron

For stars in the 15 to 20 solar mass range, fusion proceeds without a pause to form heavier and heavier elements.

As soon as a carbon core forms, it begins to contract and heat up until the fusion of carbon to form Oxygen begins at its center.

An Oxygen core then forms and contracts and heats up in turn until it starts fusing to make Neon, and so on.

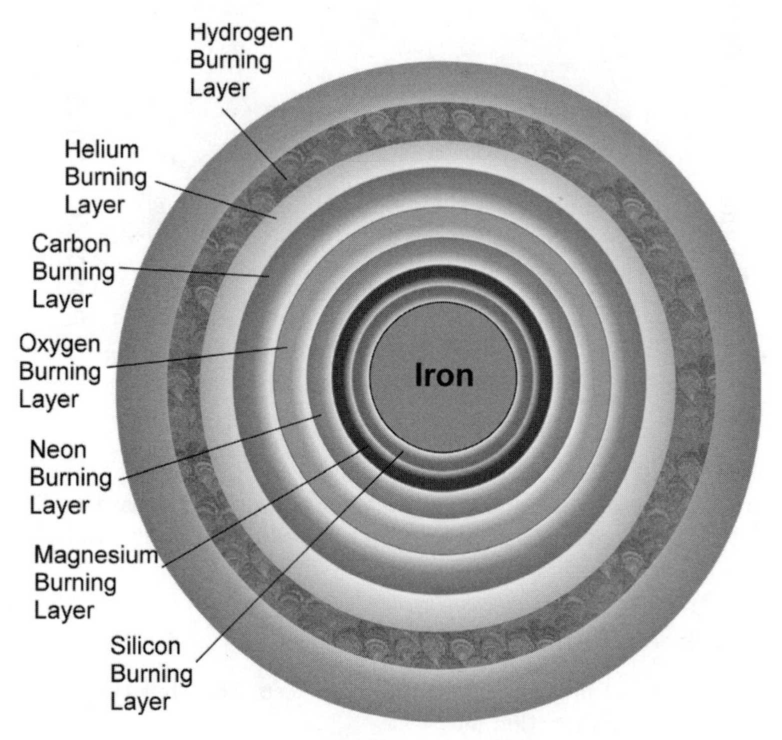

End Stage of a $20M_{\odot}$ Star

Each stage releases less energy than the last and proceeds faster.

For a 20 solar mass star, life is very short indeed:

Hydrogen burning	Lasts 10 million years.
Helium burning	Lasts 1 million years.
Carbon burning	Lasts 1000 years
Oxygen burning	Lasts 1 year
Silicon burning	Lasts 1 week

The final stage, when the iron core forms, lasts only a day.

The Story in Pictures on the HR Diagram

Unlike lower mass stars, these stars have just one red giant stage and evolve across the top the HR diagram without pausing.

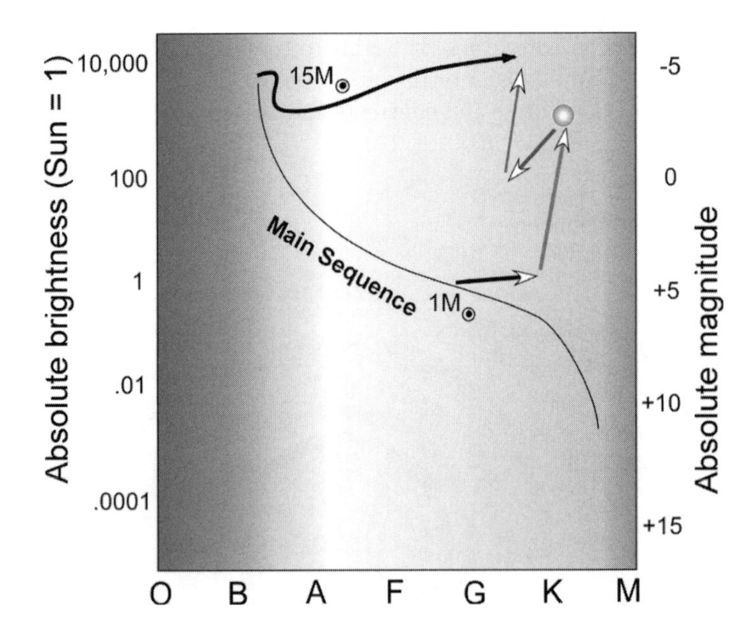

Where the one solar mass star moves vertically up and down the diagram, pausing at various stages, a 15 solar mass star moves almost horizontally, keeping constant brightness as its surface temperature drops.

027.2 Payback Time

The temperature and pressure in the iron core keep rising but there is no new energy-releasing fusion reaction left.

Instead, the temperature goes so high that it rips the iron nuclei apart and begins the formation of elements heavier than iron. Those processes absorb energy. The temperature rise then stops.

The energy pay-back has to come from somewhere. The only place it can come from is gravity and the core begins a free-fall collapse.

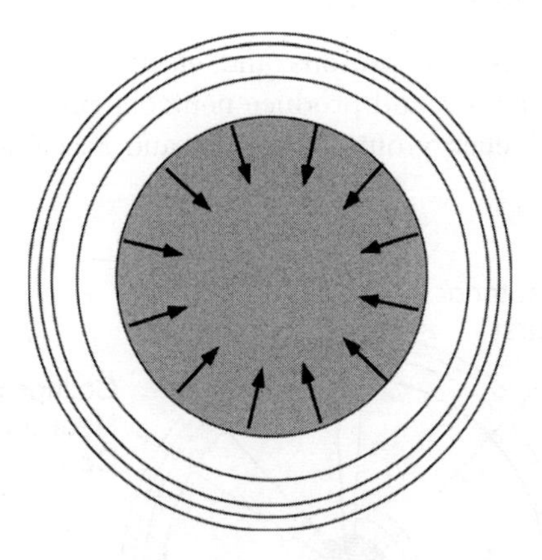

Because the temperature can no longer rise, the pressure inside the core can no longer increase. There is nothing holding the core up against its own gravity and it proceeds to collapse in on itself.

As it falls in on itself, it gains speed and energy until its energy of motion is greater than all of the energy ever released by nuclear fusion reactions in the entire history of the star.

027.3 Neutron Matter

The Electrons get squeezed out.

As the collapse proceeds, the protons and electrons react with one another through the weak interaction and produce neutrons and neutrinos.

The neutrinos carry energy out of the core and accelerate its collapse.

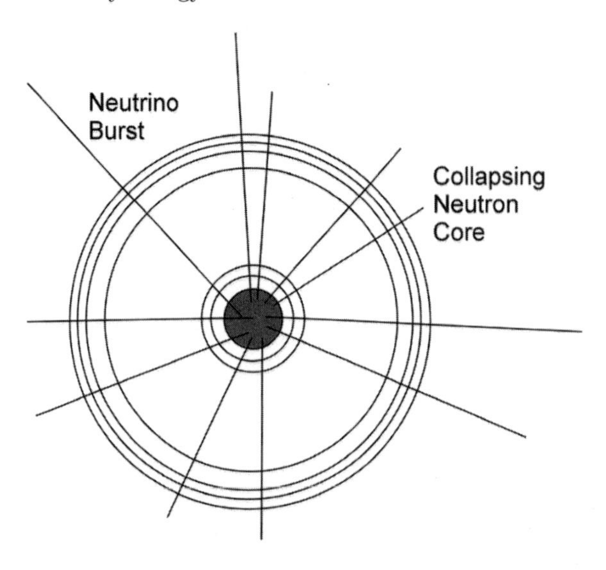

What is left is a collapsing core made of nothing but neutrons.

The electrons, which support a white dwarf by touching one another, are gone. Only neutrons, which are much smaller in size than electrons, are left. There is nothing to stop the collapse.

The Neutrino Burst

The theory predicts that each supernova should be preceded by a burst of neutrinos.

The supernova 1987a was preceded by a ten second neutrino burst that was detected by both the Kamioka II and the Irvine-Michigan-Brookhaven (Morton Salt Mine) water Cherenkov detectors. Arrays of photomultiplier tubes detected the light from neutrino interactions with the water in large acrylic vessels placed deep underground.

027.4 The Explosion

Hitting Bottom

When the collapsing neutron-matter core is just a few miles across, it is moving inward at a large fraction of the speed of light.

At that instant, the neutrons touch each other and the collapse stops.

All of the energy of motion of the falling core is instantly converted into heat and gamma rays.

More energy than the star has released in its entire lifetime is released in a fraction of a second, creating a supernova explosion.

Supernova SN1987a

The closest supernova in recent times occurred in the Large Magellanic Cloud in 1987 and was designated SN1987A. It is seen as a small hourglass figure near the center of this picture.

Here is a Hubble Space Telescope image of SN1987A. The ring is the result of the blast of ultra-violet light from the core collapse hitting gas clouds that resulted from an earlier and smaller disruption of the original star. The details of the ring, along with some blobs of material being ejected by the supernova, can be seen.

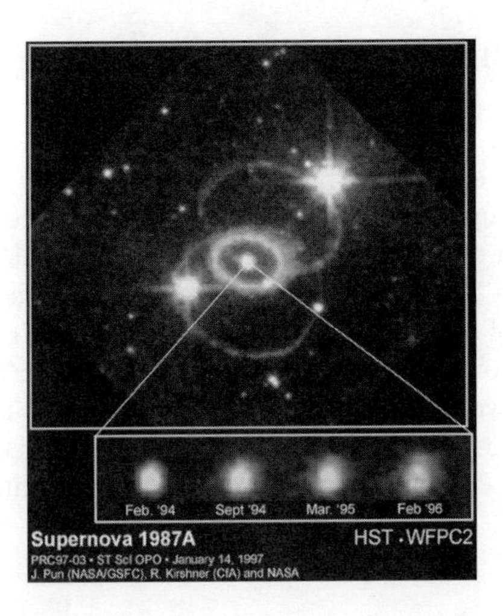

Energy Output

When a supernova occurs, it emits more light than billions of stars combined. Thus, supernovas can be seen from far away and most of them are extremely far away. Here is a supernova seen in 2006 in the galaxy M100. The picture was taken by the Ultraviolet/Optical Telescope after a gamma-ray burst was detected by the Burst Alert Telescope operated by NASA's Swift Science Center.

Origin of the Heavy Elements

All of the elements heavier than iron were formed and blown out into interstellar space during supernova explosions. Nuclear fission of uranium actually taps the stored energy of ancient supernovas.

027.5 Types: I Clean and II Dirty

Classification

There are two types of supernovas.

The simple and straightforward type occurs when the core of a giant star collapses, blowing the star to bits. While logic might dictate otherwise, astronomers have labeled this Type II. SN1987A is a type II supernova.

The indirect type of supernova occurs when a medium mass star evolves to a large white dwarf that orbits close to a normal star. Matter from the normal star falls onto the white dwarf, slowly increasing its mass. When its mass exceeds 1.4 solar masses, the electrons that are holding the dwarf up combine with the protons and a collapse begins. This delayed type of supernova is called Type I.

Type II

The signature of a Type II supernova is that it lasts for weeks because it takes a long time for the blast to work its way out from inside the giant star. Because the star still had a hydrogen outer envelope when it exploded, hydrogen lines are seen in the spectrum of the supernova. Because of all the material in the way of the blast, this type could be called a "Dirty Supernova."

Type I

The signature of a Type I supernova is that it is usually very short and shows no hydrogen lines at all. It could be called a "Clean Supernova."

Type I supernovas are extremely important because the core is always the same size when it starts its collapse and the collapse always happens in the same way, so these are standard light sources.

Because they are also among the brightest light sources in the universe, they can be identified from great distances.

027 Spot Check

Here are some questions to check your understanding of the material in module 027. Both the answers and where to find these questions at the website may found at the end of the Study Guide.

1 When the iron nuclei in the core of an evolved high-mass star start to come apart, they

a. trigger a new round of nuclear fusion.

b. absorb energy and limit the core temperature.

c. release energy and raise the core temperature.

d. cause the core to expand.

2 Elements with more protons and neutrons than iron (Uranium for example), are believed to have formed

a. during supernova explosions of very massive stars.

b. during novas caused by white dwarfs that steal fuel from their neighbors.

c. during the Big Bang that began the universe.

d. through fusion reactions in stars similar to our Sun.

3 Near the end of the life of a massive star, an intense burst of neutrinos

a. happens when iron nuclei are torn apart.

b. is not likely to happen.

c. happens after the neutron core forms.

d. happens when the electrons are eliminated.

4 Type I supernovas have the following properties:

a. a spectrum with no hydrogen lines and a standard maximum brightness.

b. a spectrum with hydrogen lines and a standard maximum brightness.

c. a spectrum with no hydrogen lines and a variable maximum brightness.

d. a spectrum with hydrogen lines and a variable maximum brightness.

5 Our Sun is expected to last for about 10 billion years. Which of these is a likely lifespan for a star with 20 times the mass of our Sun?

a. 1 million years.

b. 10–15 million years.

c. 200 billion years.

d. 20–30 billion years.

e. 0.5–1 billion years.

028: Collapsed Objects

028.1 The Pulsar in the Crab Nebula

Discovery

Here we have the standard science fiction movie scenario: Late one night in 1967, Susan Jocelyn Bell, a young graduate student (Physics BS degree from University of Glasgow) was operating a radio telescope that she had helped to construct. Regular signals appeared on the chart recorder. The signal was repeating over and over, thirty times a second. She had discovered the first "*pulsar.*"

The Little Green Men Theory

The LGM theory was that the signal was an interstellar "standard time" signal beeping away at regular intervals. In this case, thirty times a second.

That theory did not last long because it was soon observed that the signal is slowing down. There would be no point to a standard time beacon that did not keep good time.

Tommy Gold's Crazy Idea

The radio signal was identified as coming from a supernova remnant, the Crab Nebula, which suggested an idea to Tommy Gold. He sort of specialized in crazy ideas that turn out to have been right.

Tommy Gold's crazy idea about pulsars was that the repeating radio signals were coming from the collapsed neutron core left behind by a supernova explosion. That was a crazy idea because the idea of neutron matter was purely theoretical and nobody really believed that a supernova explosion would leave anything behind at all.

His reasoning went like this:

In a type II supernova, much of the energy of the explosion is diverted to the outer layers of the star and the collapsed core can stay intact, as a *neutron star*.

The collapse intensifies the magnetic field that was trapped in the star and also increases the rate of spin.

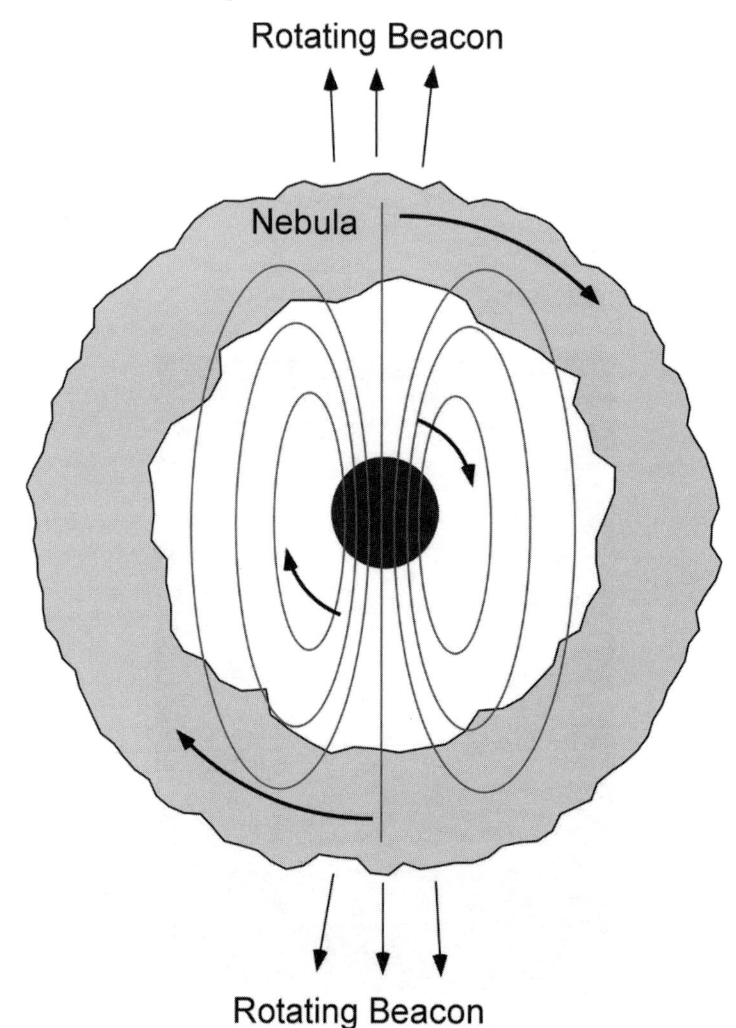

A spinning magnet just a few miles across but with a mass similar to the mass of our sun is rotating thirty times a second and dragging its magnetic field through the surrounding nebula.

As the neutron star spins, it drags its intense magnetic field through the surrounding nebula, generating a pair of rotating beams of electromagnetic radiation.

Whenever one of the rotating beams swings past the earth, we see a pulse: Thirty pulses per second.

Gold was right.

Gold's theory predicted that the dragging magnetic field would slow the spin of the neutron star and transfer energy from to the surrounding nebula. The signal was indeed observed to be slowing down and the rate at which the Crab pulsar is losing energy is essentially the same as the rate at which the nebula is shining, so the spinning neutron star is acting as a dynamo to light up the nebula.

Since the discovery of the Crab pulsar, hundreds of others have been found. Some, like the pulsar in the Vela supernova remnant (shown on our website) are much older.

Thanks to Tommy Gold's 'crazy idea,' we now realize that all of these pulsars are actually spinning *neutron stars.*

028.2 X-Ray Sources

How the X-rays Are Produced

When matter falls onto a neutron star, a significant fraction of its mass is converted into energy simply because it is falling in an intense gravitational field.

When a neutron star has a normal star for a partner, gas from the normal star spirals in to the neutron star, emitting a steady stream of X-rays.

When enough nuclear fuel builds up on the surface of the neutron star, a short fusion reaction starts and emits a burst of X-rays until the fuel is gone. The result is called an *X-ray burster.* Notice that this process is *exactly the same* as the one that produces a nova when it happens to a White Dwarf star. Here the collapsed object is much smaller, the gravitational field is much stronger, and the energies are much higher.

Observations of X-rays

There are a lot of X-ray sources out there. Our website shows an all-sky map of the sources seen by the ROSAT X-ray observatory. It is omitted here because it relies on color and does not work in black and white. The largest number of these sources are neutron stars in binary systems.

Here is a graph of the X-ray intensity from a typical X-ray burster.

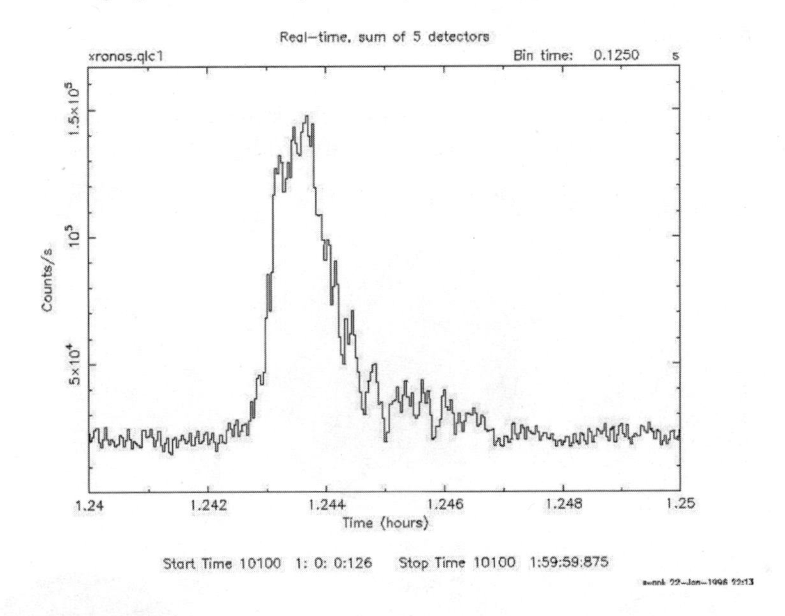

Notice that the whole graph shows only one hundreth of an hour — 6/10 of a minute of data.

028.3 Black Holes

The Idea of the Event Horizon

Light falls in a gravitational field, just like everything else does.

As the mass of a neutron star increases, the strength of gravity near its surface becomes enough to prevent even an outward light ray from escaping.

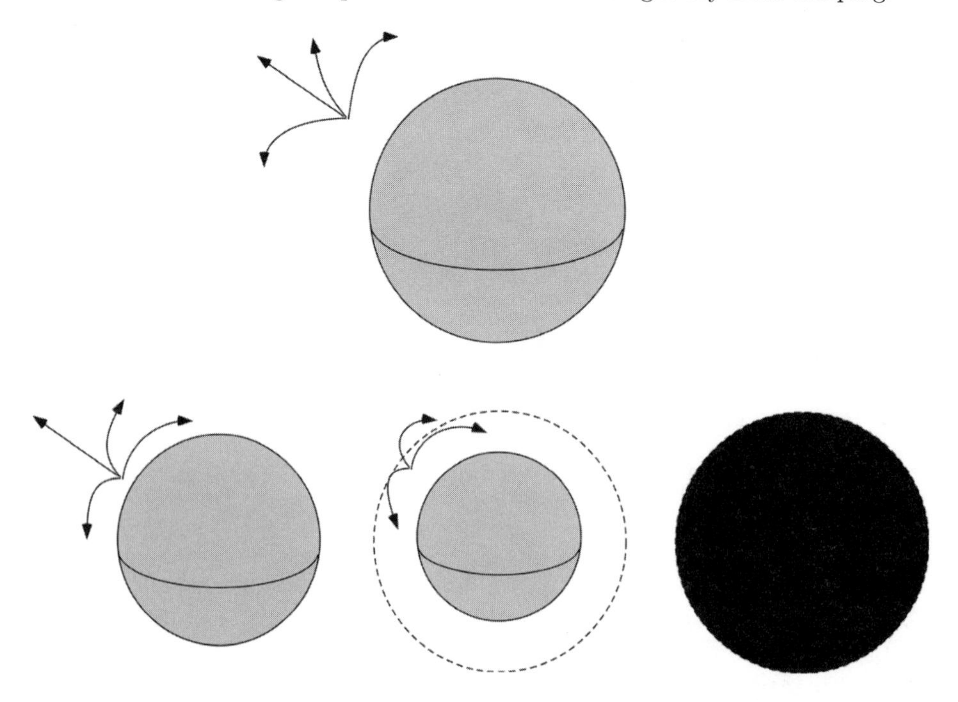

Because nothing can travel faster than light, nothing else can escape either and a 'point of no return' or event horizon develops around the star.

Because the event horizon absorbs all light (and anything else) that comes near it, it is called a *black hole.*

Time Distortion

The event horizon is a property of space and time near the collapsing star.

Once the horizon forms, it no longer matters what the star is made out of. No signal from it can ever reach us again.

The nature of time near the horizon is altered. A clock that is dropped in to the black hole will be seen to run slower and slower as it approaches the horizon. From the outside, we will never see it enter at all. However, the light from the falling clock will be shifted to longer and longer wavelengths until we can no longer see it.

Similarly, if we watch a neutron star collapse to a black hole, the last signals from the star will simply change ever more slowly as they shift to longer and longer wavelengths. The surface of the star will seem to 'freeze' at the newly forming horizon. For this reason, Soviet scientists insisted on referring to these objects as 'frozen stars'.

Only Slightly Crazier than Neutron Stars

While a neutron star is much smaller than the next larger collapsed object, a white dwarf star, there is not that much difference between a one solar mass neutron star and a black hole of just a few solar masses.

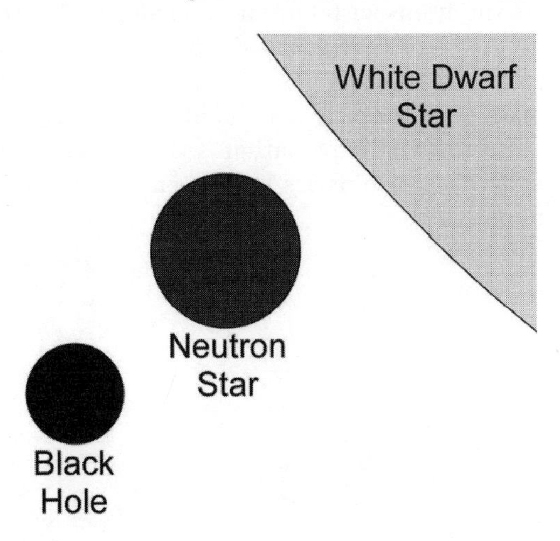

Gravity theorists suspected (beginning in 1939 with papers by Oppenheimer and his coworkers Volkoff and Snyder) that it would be very difficult for Nature to form neutron stars without occasionally going too far and making a black hole.

For many years, both neutron stars and black holes were regarded as crazy predictions. Once it was realized that neutrons stars exist in abundance, black holes did not seem so crazy any more.

028.4 Detecting Black Holes

The Prediction

It was predicted (by a Russian scientist, Ya B. Zeldovich) that a black hole should generate a characteristic fluctuating X-ray signal.

Because so many stars are in binary pairs, a black hole might be paired with an ordinary star. Matter from the ordinary star would stream toward the black hole and gain enormous speed as it approached the event horizon.

Because the event horizon is only a few miles across and gas is coming from a region the size of our Sun, there would be a "traffic jam" outside the hole.

Gas spiralling in to the hole would hit the "traffic jam" and heat up, emitting X-rays.

The Test

The very first X-ray detectors that were sent above the atmosphere on rockets in the seventies detected something like the predicted signature of a black hole in the constellation Cygnus. The first X-ray source seen in Cygnus was named Cygnus X-1.

The source flickered in just hundreths of a second, which was the right time scale for a black hole binary.

By correlating its variations with known radio and optical sources, its binary partner was identified (HDE 226868). The doppler shift of the partner's spectral lines revealed the orbit and thus the possible mass of the invisible X-ray source.

At a probable mass of 7 times the mass of our sun, it was far too massive to be a neutron star. The simplest explanation is that it is a black hole.

About ten other black hole binaries have been discovered, so there is little doubt that they are out there.

028 Spot Check

Here are some questions to check your understanding of the material in module 028. Both the answers and where to find these questions at the website may found at the end of the Study Guide.

1 One conclusion that was drawn from the gradual slowing of the radio signals from the Crab Nebula was that they were probably

a. of artificial origin.

b. from a source moving away from us.

c. of natural origin.

d. an obvious hoax.

e. from a source moving toward us.

2 A black hole that has formed from the collapse of a star is expected to be

a. similar in size to a neutron star.

b. less than 1/10 the size of a neutron star.

c. 100 times the size of a neutron star.

3 Black holes

a. have been detected as pulsing radio sources.

b. have been detected because they block starlight.

c. have been detected because infalling matter emits X-rays.

d. cannot be detected because they emit no radiation.

4 The point at which even an outwardly directed light ray is pulled into a black hole is called the

a. tipping point.

b. event horizon.

c. particle horizon.

d. critical point.

029: The Milky Way Galaxy

029.1 How we see it

In the night sky

The Milky Way: On a clear night, far from the city lights, a band of glowing clouds stretches across the sky.

Galileo noticed that they are not clouds. They are stars, hundreds of billions of them. The thickest part of the Milky Way is near the constellation Sagittarius.

On the Celestial Sphere

On the Celestial Sphere, the Milky Way follows a great circle path that is different from both the Celestial Equator and the Ecliptic.

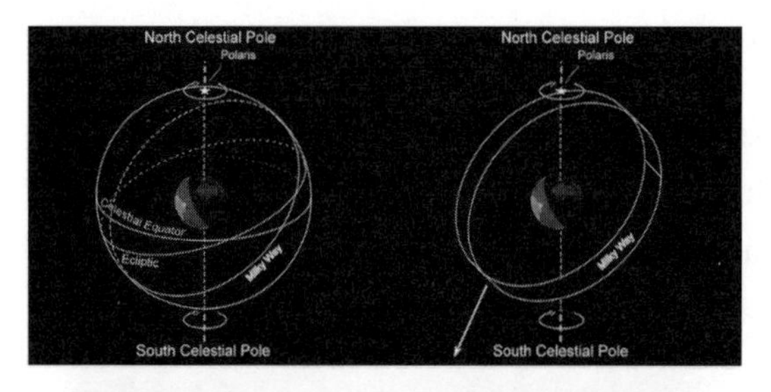

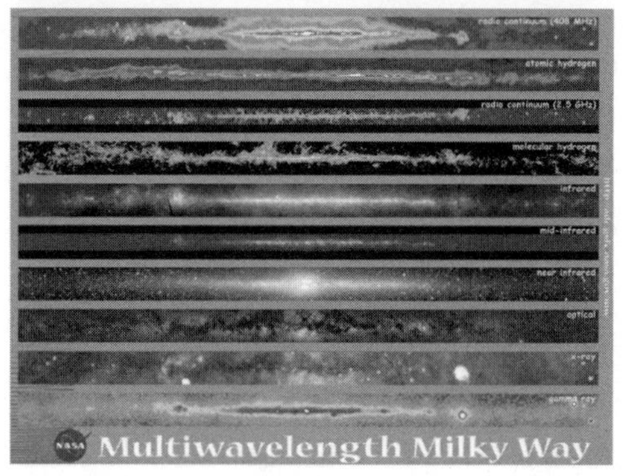

A narrow strip of sky contains most of the stars that belong to the Milky Way.

029.2 How we Measure Distances Within It

Cepheid Variable Stars

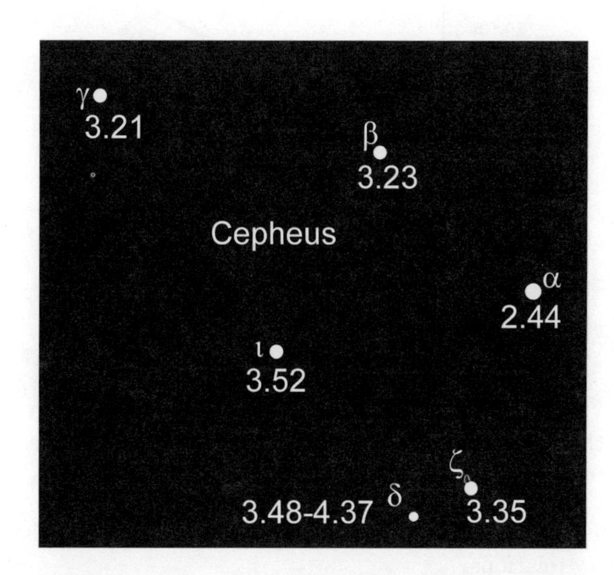

Cepheid variable stars, named after the first one discovered, Delta Cepheus shown here, vary in brightness with regular periods.

Cepheids are good distance markers because their luminosities are simply related to their periods. Because they are high-mass *supergiant* stars, they are *very bright and very rare.*

Cepheid star periods range from 3 to 50 days. The longer the period, the more massive and brighter the star.

Delta Cepheus cycles between apparent magnitudes of 3.6 and 4.3 with a period of 5.4 days.

Polaris is also a Cepheid variable but only changes from apparent magnitude 2.5 to apparent magnitude 2.6.

RR Lyra Variable Stars

RR Lyrae variable stars are also good distance markers. They all have similar periods and similar luminosities.

RR Lyra star periods are usually between 0.3 and 0.7 days.

Because they are actually *horizontal branch stars*, they are *dimmer* than Cepheids but *much more numerous.*

Using Variable Stars as Standard Candles

To find the distance to a distant collection of stars:

1. Measure the distances and apparent magnitudes of nearby variable stars and calculate their absolute magnitudes or lumonosities. The result is

called the *period-luminosity* relation. First done by Henrietta Swan Leavitt in 1912, for 25 variable stars in the Small Magellanic Cloud.

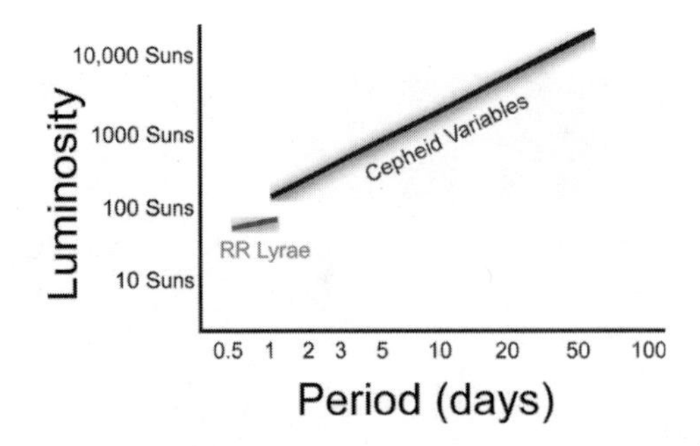

2. Measure the periods of variable stars in the distant collection and infer their absolute magnitudes from the period-luminosity relation.

3. Measure the apparent magnitudes of the distant variable stars and calculate their distance.

One *important problem*: Cepheids and RR Lyra variables have different period-luminosity relations.

029.3 Where are We in it?

Relative to the Stars We Can See

Because of dust and gas in the plane of the Milky Way, we can see only part of it. The part that we can see is about 6000 light years thick and 30,000 light years across.

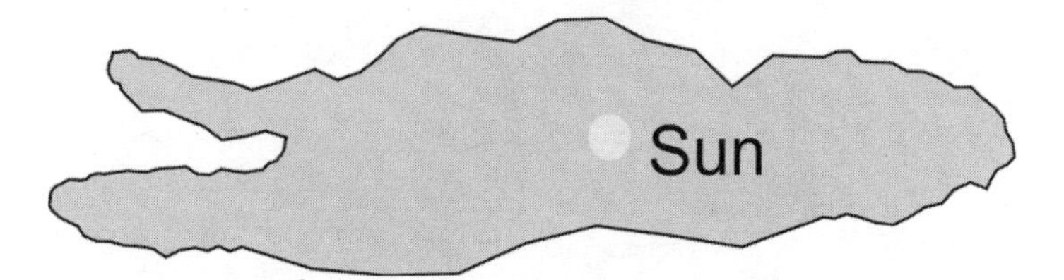

Naturally, we are in the center of what we can see.

Globular Clusters

Globular Clusters are made up of millions of stars and are at about the same distance from us as the stars of the Milky Way. The one shown here is PIA10372, the largest and brightest of the globular clusters orbiting our galaxy.

The globular clusters are near our galaxy but many are out of its plane, so we can see them far away. They act like tall buildings in a city. By looking up at them, we can find out where in the city we are.

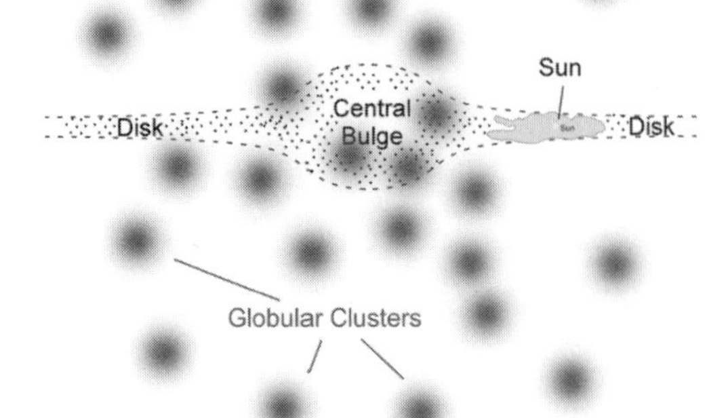

The Copernican Principal

If we take the globular clusters as markers of the overall size of our galaxy, we get this general picture:

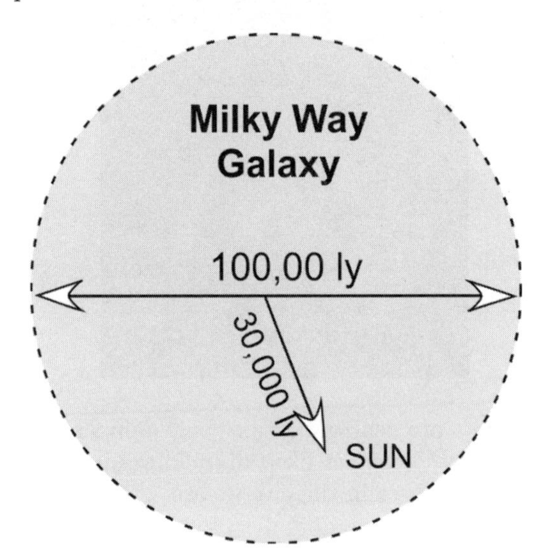

We are about 30,000 light years from the center, about 3/5 of the way out in a galaxy that is 100,000 light years across.

The Copernican Principal is: *We are never special.* We are not at the center of our earth. We are not at the center of the solar system. Our solar system is not at the center of the Milky Way.

029.4 What is its Overall Shape?

Clues from the Extragalactic Nebulae

Astronomers noticed many cloud-like objects that were clearly outside of our own galaxy and called them "Extragalactic Nebulae." Variable stars in these objects were used to determine that they were distant enough to be outside out galaxy.

The original distance results indicated that these objects were close to our galaxy and much smaller than our galaxy. Those distances were wrong because RR Lyra variables had been confused with Cepheid variables.

Once the true distance to the "extragalactic nebulae" became known, it was realized that they are actually other galaxies like our own.

The galaxies that seem to be most like ours are spiral shaped like our neighbor, Andromeda

Another galaxy similar to our own is M51, with its spiral arms on display.

Mapping the Milky Way

By using the full set of electromagnetic wavelengths, we can now probe through the dust and gas and get a better picture of our own Milky Way Galaxy.

The resulting map looks like this sketch.

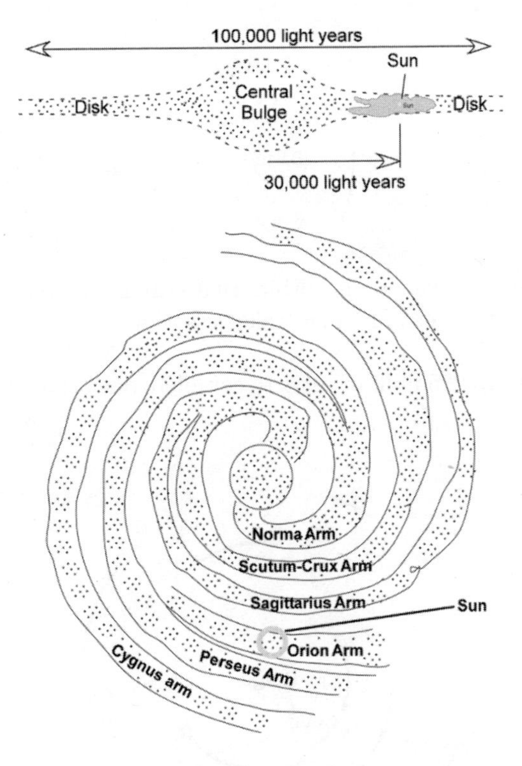

Some of the spiral arms have been mapped out. Our Sun is in what is called the "Orion Arm".

The spiral arms stand out because they contain many bright, massive, type O and B stars.

Less massive stars, such as our sun, are distributed uniformly all through the disk and orbit in and out of the spiral arms.

Where the Stars Are

The traditional classification of stars in our galaxy is:

Population I: Stars closely related to our Sun.

Population II: Very old stars not much like our Sun.

Population I stars are mostly found in the disk and include many short-lived bright type O and B stars as well middle-aged stars such as our sun.

Population II stars are mostly found in the halo and are all extremely old.

As usual, the traditional terminology is confusingly backwards: Population II came before Population I.

Halo stars and globular clusters have random orbits around the central bulge. Their orbits are inclined at all angles and can be elliptical. They occasionally pass through the disk but cross it quickly and keep going.

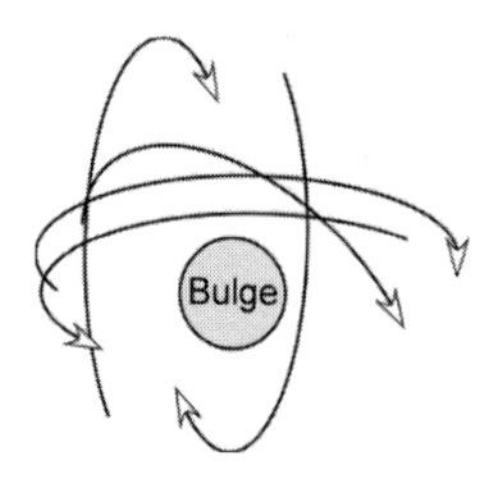

Population I stars have circular orbits that stay in the plane of the disk.

Stars closer to the central bulge take less time to go around than stars farther out.

Our own sun takes 225 million years to complete one orbit around the central bulge.

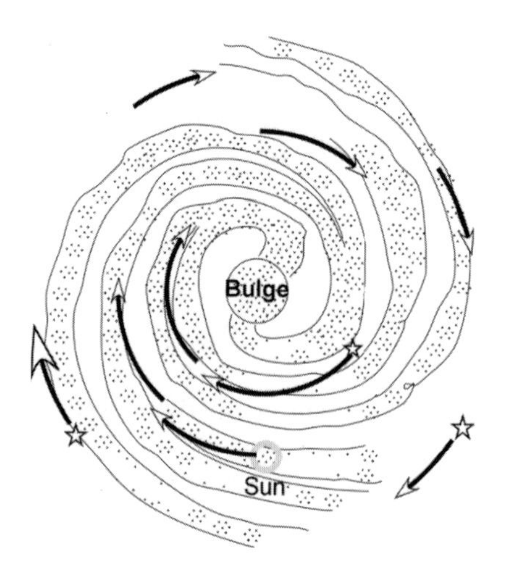

029.5 The Monster in the Core

Star Swarm

Near the center of the Milky Way, stars are packed tightly together so that collisions are quite common.

The average distance between stars is about 1/100 of what it is near our Sun. That would put the nearest stars well inside of our Oort clo

Closer stars would appear brighter, 10,000 times as bright as they do here. The very brightest stars in the sky would be as bright as our full moon.

The night sky would be a million times brighter than it is here. Infra-red telescopes, some on the ground and some in orbit, have penetrated the dust and given pictures of the stars at the galactic center.

A Powerful Source at the Core

Powerful sources of infra-red light, X-rays and radio waves are at the center of our galaxy. This picture is an X-ray image with the constant-intensity contours of the radio signal superimposed on it.

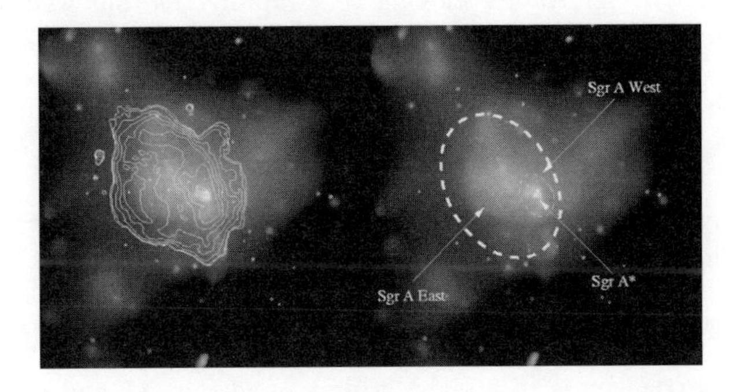

The radio contours pick out a "hot spot" that has been named Sgr A*.

The abbreviation Sgr stands for Sagittarius and is usually pronounced "saj".

The source Sgr A* has a luminosity that is a million times the luminosity of our sun.

Here is a more recent (Jan 6, 2003) and much more detailed X-ray image of this region:

Star Motions Near Sgr A*

The stars closest to Sgr A* are moving very rapidly. On our website you can see that motion in some animated infrared images, taken from 1992 to 1998.

These velocities can be measured and provide a mass for the central object that these stars orbit.

From each of the stars in orbit around Sgr A*, the mass inside of that orbit can be found.

All of the mass, 2.9 million times the mass of our sun, is concentrated within a region smaller than 0.01 parsecs across.

Sgr A* is a Monster Black Hole

One star, in particular, is in a tight orbit around Sgr A* and we have been able to watch it complete almost one full orbit. The orbit is only two light-days across. Only one kind of object could pack almost 3 million solar masses into such a small space. There is no longer any real doubt that Sgr A* is a black hole.

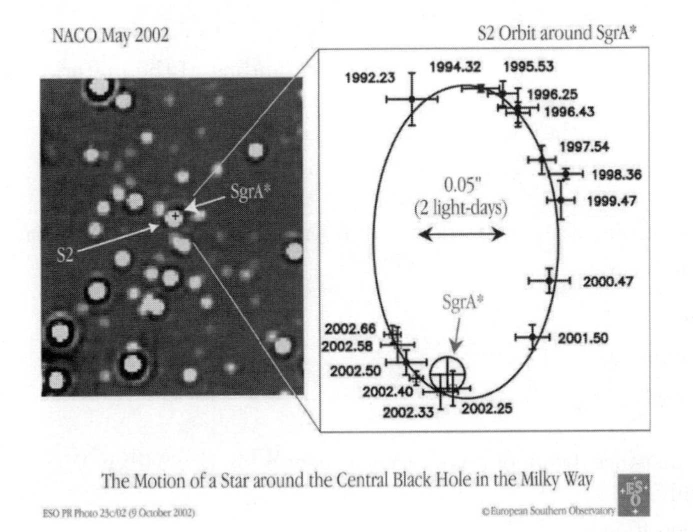

The Motion of a Star around the Central Black Hole in the Milky Way

The hole is about 4.5 million kilometers in diameter — somewhat larger than our Sun.

For more detail about the black hole and an impressive movie of stars zinging around the hole, go to Galactic Research at MPE (http://www.mpe.mpg.de/ir/GC/index.php). The site will open in a new window. Note that the animated GIF takes a while to load and you can come back here by just closing the window.

Just as with smaller black holes, the radio waves, infra-red light, and X-rays that we see from Sgr A* come from matter that is falling toward the hole. The lobes of hot gas that are seen near Sgr A are exactly what would be predicted from gas falling into such a black hole.

029 Spot Check

Here are some questions to check your understanding of the material in module 029. Both the answers and where to find these questions at the website may found at the end of the Study Guide.

1 The motions of the stars near Sgr A* make it possible to determine its

a. temperature.

b. size.

c. luminosity.

d. mass.

2 The distance from our sun to the center of the Milky Way Galaxy is roughly

a. 6000 light years.

b. 100,000 light years.

c. 300,000 light years.

d. 1000 light years.

e. 30,000 light years.

3 Our Sun's location in the Milky Way is closest to the

a. Cygnus Arm.

b. Sagittarius Arm.

c. Orion Arm.

d. Scutum-Crux Arm.

e. Norma Arm.

4 The constellation Sagittarius is where the Milky Way

a. is thickest.

b. has its most northern point.

c. cannot be found.

d. is thinnest.

e. splits into two bands.

5 The source that is called Sgr A* emits

a. both infrared light and radio waves.

b. only infrared light.

c. only radio waves.

6 Extragalactic nebulae such as the Great Nebula in Andromeda were not immediately recognized as galaxies similar to our own Milky Way because they were thought to be

a. smaller, nearby objects.

b. the wrong shape.

c. larger, very distant objects.

7 In comparison to RR Lyra variables, Cepheid variable stars are

a. less luminous and more common.

b. more luminous and more common.

c. more luminous and less common.

d. less luminous and less common.

Answers to Spot-Check Questions

All of the spot-check questions are taken from the course website at

http://www.people.vcu.edu/~rgowdy/

Here you can find the correct answers to the questions and also where to look for them in the website. For example the reference "Module 002.203" tells you that the question is the third question in part 2 of module 002. In that case, the website link to the question will be just 002.203. The website versions of the questions have active links that explain each answer choice.

Module 001

Answers

1 `Choice d.` (Joe's observations were not reproduced.)

2 `Choice d.` (only south of the arctic circle and north of the antarctic circle.)

3 `Choice a.` (that everyone on Earth sees the Sun rise and set each day.)

4 `Choice a.` (I saw the Moon rise at 6:52pm yesterday.)

5 `Choice a.` (constellations that could be seen from Egypt but not from Greece.)

Where to find these questions in the website

1 Module 001.202

2 Module 001.303

3 Module 001.401

4 Module 001.103

5 Module 001.503

Module 002

Answers

1 `Choice a.` (what holds up the surface of the Earth.)

2 `Choice d.` (well-tested models.)

3 `Choice c.` (noon Sun angles at two locations to determine the angle between the Earth radii to those locations.)

4 `Choice b.` (southern constellations were seen higher in the sky in Egypt than in Greece.)

5 `Choice b.` (a bad thing since it lets the model fit many possible measurements.)

Where to find these questions in the website

1 Module 002.203

2 Module 002.501

3 Module 002.403

4 Module 002.302

5 Module 002.104

Module 003

Answers

1 `Choice c.` (1:59:00 am the next day.)

2 `Choice e.` (a constellation.)

3 `Choice c.` (a few minutes shorter than a solar day.)

4 `Choice e.` (moves eastward along the ecliptic by 1°.)

5 `Choice b.` (the noon sun is highest in the sky.)

6 `Choice a.` (a sidereal day.)

Where to find these questions in the website

1 Module 003.304

2 Module 003.202

3 Module 003.402

4 Module 003.504

5 Module 003.601

6 Module 003.104

Module 004

Answers

1 `Choice e.` (The retrograde motion of the planets.)

2 `Choice b.` (it did not account for observations any better than the Ptolemaic System)

3 `Choice b.` (westward motion of the planets relative to the stars.)

4 `Choice a.` (did not really have anything fixed in place at the center of the universe.)

5 `Choice d.` (even the closest stars are so far away that the shift is very small.)

Where to find these questions in the website

1 Module 004.502

2 Module 004.604

3 Module 004.102

4 Module 004.401

5 Module 004.303

Module 005

Answers

1 `Choice a.` (None of these systems.)

2 `Choice c.` (Kepler.)

3 `Choice b.` (1/60 degree of arc.)

4 `Choice d.` (it used circular orbits instead of ellipses.)

5 `Choice a.` (move slower on the average.)

Where to find these questions in the website

1 Module 005.302

2 Module 005.502

3 Module 005.204

4 Module 005.402

5 Module 005.603

Module 006

Answers

1 `Choice a.` (There is great disinterest because there is no need to replace a theory that has passed every observational test. Nobody at all comes to Fred's talk.)

2 `Choice b.` (There are fish in Lake Nyak.)

3 `Choice c.` (The Moon is made entirely of cheese.)

4 `Choice a.` (subject to change.)

Where to find these questions in the website

1 Module 006.301

2 Module 006.204

3 Module 006.101

4 Module 006.401

Module 007

Answers

1 `Choice b.` (the Sun is farther from the Earth and larger than the Moon.)

2 `Choice c.` (Waxing gibbous.)

3 `Choice b.` (A simulation, using currently accepted physical laws, of waves crashing on the beach.)

4 `Choice c.` (counterclockwise.)

5 `Choice d.` (Someone in the Netherlands.)

6 `Choice b.` (the quarter phases of the Moon.)

Where to find these questions in the website

1 Module 007.201

2 Module 007.508

3 Module 007.102

4 Module 007.401

5 Module 007.604

6 Module 007.301

Module 008

Answers

1 `Choice c.` (the bullet.)

2 `Choice c.` (its acceleration will be zero.)

3 `Choice d.` (build things that he could measure.)

4 `Choice d.` (10 Newtons.)

5 `Choice c.` (sometimes the wooden ball hit first, sometimes the iron one hit first.)

6 `Choice d.` (slow down and stop.)

Where to find these questions in the website

1 Module 008.604

2 Module 008.304

3 Module 008.203

4 Module 008.508

5 Module 008.403

6 Module 008.101

Module 009

Answers

1 `Choice c.` (to every other object in the universe.)

2 `Choice c.` (toward the Earth.)

3 `Choice b.` (Increase its speed to 6 miles per second and then give it another speed boost when its distance from the Earth stops increasing.)

4 `Choice e.` (how planets move and how the tides work.)

5 `Choice c.` (almost the same motions but with corrections.)

Where to find these questions in the website

1 Module 009.301

2 Module 009.201

3 Module 009.602

4 Module 009.404

5 Module 009.503

Module 010

Answers

1 `Choice d.` (Neptune)

2 `Choice c.` (Venus, with no moons at all.)

3 `Choice b.` (rock and possibly iron.)

4 `Choice a.` (Jovian Planet.)

5 `Choice c.` (one of the larger dwarf planets in the Kuiper Belt.)

6 `Choice c.` (comet.)

Where to find these questions in the website

1 Module 010.103

2 Module 010.204

3 Module 010.401

4 Module 010.304

5 Module 010.603

6 Module 010.504

Module 011

Answers

1 `Choice e.` (the planet Mars.)

2 `Choice d.` (is elliptical enough to give us an annular solar eclipse when the Moon is near its apogee.)

3 `Choice d.` (Mostly as ice and water vapor.)

4 `Choice b.` (hydrated minerals.)

5 `Choice b.` (impossible because every part of it is too hot for water ice.)

6 `Choice b.` (about 20 successful space probes.)

7 `Choice e.` (Mars)

8 `Choice b.` (Mariner 10.)

9 `Choice c.` (rotates backwards so that the Sun rises in the West.)

10 `Choice d.` (elliptical enough to make the intensity of sunlight vary by 40 percent.)

11 `Choice c.` (maria.)

12 `Choice c.` (elliptical enough to make the intensity of sunlight vary by 6 percent.)

13 `Choice d.` (1969.)

14 `Choice b.` (Mercury)

Where to find these questions in the website

1 Module 011.520

2 Module 011.411

3 Module 011.505

4 Module 011.416

5 Module 011.203

6 Module 011.214

7 Module 011.514

8 Module 011.115

9 Module 011.205

10 Module 011.509

11 Module 011.402

12 Module 011.303

13 Module 011.423

14 Module 011.102

Module 012

Answers

1 `Choice c.` (Jupiter)

2 `Choice a.` (an atmosphere of Hydrogen and Helium with some methane.)

3 `Choice c.` (an atmosphere of Hydrogen and Helium with some methane.)

4 `Choice c.` (Titan)

5 `Choice e.` (an atmosphere of Hydrogen and Helium with no real surface.)

6 `Choice d.` (Cassini-Huygens)

7 `Choice e.` (Uranus)

8 `Choice a.` (displaced from its rotation axis poles and also from the center of the planet.)

9 `Choice b.` (Saturn)

10 `Choice e.` (Jupiter)

11 `Choice e.` (Voyager 2)

12 `Choice c.` (Jupiter)

13 `Choice a.` (Jupiter)

14 `Choice d.` (17 hours.)

15 `Choice c.` (Neptune)

Where to find these questions in the website

1 Module 012.112

2 Module 012.301

3 Module 012.401

4 Module 012.215

5 Module 012.201

6 Module 012.225

7 Module 012.318

8 Module 012.311

9 Module 012.210

10 Module 012.116

11 Module 012.411

12 Module 012.124

13 Module 012.108

14 Module 012.307

15 Module 012.408

Module 013

Answers

1 `Choice c.` (at least a quarter of the way to the nearest star.)

2 `Choice d.` (meteorite.)

3 `Choice d.` (beyond the Kuiper Belt.)

4 `Choice b.` (away from the Sun.)

5 `Choice a.` (beyond all of the Jovian planets.)

Where to find these questions in the website

1 Module 013.501

2 Module 013.201

3 Module 013.401

4 Module 013.101

5 Module 013.306

Module 014

Answers

1 `Choice a.` (all of them near the ecliptic.)

2 `Choice c.` (it was too hot for volatile gases and water to condense.)

3 `Choice d.` (The Sun's final collapse blew all the gas away.)

4 `Choice d.` (far brighter than typical stars because of their large surface area.)

5 `Choice d.` (icy objects condensed out just beyond Neptune.)

Where to find these questions in the website

1 Module 014.106

2 Module 014.301

3 Module 014.401

4 Module 014.203

5 Module 014.506

Module 015

Answers

1 `Choice b.` (semiliquid rock.)

2 `Choice c.` (Greenhouse Effect.)

3 `Choice c.` (much less than the typical amount of water.)

4 `Choice c.` (both solids and liquids.)

5 `Choice e.` (sinks.)

6 `Choice c.` (absorption of ultraviolet radiation.)

Where to find these questions in the website

1 Module 015.503

2 Module 015.404

3 Module 015.102

4 Module 015.509

5 Module 015.202

6 Module 015.306

Module 016

Answers

1 `Choice b.` (there is no liquid water there.)

2 `Choice c.` (returned to the atmosphere by volcanos when the sea floor is pulled deep into the Earth.)

3 `Choice c.` (a descending convection current in the Earth's mantle.)

4 `Choice c.` (slipping tectonic plates.)

Where to find these questions in the website

1 Module 016.501

2 Module 016.401

3 Module 016.106

4 Module 016.204

Module 017

Answers

1 `Choice c.` (are kicked out of the asteroid belt by Jupiter's gravity.)

2 `Choice d.` (think about ways to get off the planet.)

3 `Choice c.` (iron and nickel.)

4 `Choice a.` (blocking the sunlight with its smoke and dust.)

Where to find these questions in the website

1 Module 017.101

2 Module 017.404

3 Module 017.203

4 Module 017.304

Module 018

Answers

1 `Choice d.` (carbon forms more stable compounds than silicon.)

2 `Choice e.` (DNA.)

3 `Choice b.` (the attraction between the hydrogen atoms on one water molecule and the oppositely charged oxygen atom on another.)

4 `Choice c.` (sunlight.)

5 `Choice a.` (microwaves.)

6 `Choice e.` (frequencies or colors present in a light source.)

7 `Choice a.` (4 times as high.)

8 `Choice a.` (dissolves in liquid water.)

Where to find these questions in the website

1 Module 018.203

2 Module 018.504

3 Module 018.305

4 Module 018.404

5 Module 018.102

6 Module 018.116

7 Module 018.112

8 Module 018.310

Module 019

Answers

1 `Choice c.` (Mars has water frozen in its ice caps and may have liquid water below its surface.)

2 `Choice c.` (habitable zone.)

3 `Choice b.` (expected lifetime of a communication-capable civilization.)

4 `Choice a.` (patterns of cracks in the ice on its surface.)

5 `Choice e.` (Alpha Centauri.)

6 `Choice c.` (are too small.)

Where to find these questions in the website

1 Module 019.203

2 Module 019.402

3 Module 019.506

4 Module 019.305

5 Module 019.412

6 Module 019.209

Module 020

Answers

1 `Choice a.` (1036 seconds of arc.)

2 `Choice c.` (Cassegrain Focus.)

3 `Choice c.` (4 parsecs.)

4 `Choice c.` (0.1 seconds of arc.)

5 `Choice c.` (a shallow bowl with the open part facing up.)

Where to find these questions in the website

1 Module 020.501

2 Module 020.202

3 Module 020.405

4 Module 020.401

5 Module 020.103

Module 021

Answers

1 Choice e. (the masses and diameters of both stars in the system.)

2 Choice d. (1200 m/s)

3 Choice e. (two regions of maximum pressure.)

4 Choice b. (a visual binary system.)

5 Choice e. (the distance traveled by a crest divided by the time taken.)

6 Choice b. (moving away from us.)

Where to find these questions in the website

1 Module 021.307

2 Module 021.110

3 Module 021.102

4 Module 021.304

5 Module 021.107

6 Module 021.201

Module 022

Answers

1 `Choice b.` (11.)

2 `Choice c.` (100 parsecs.)

3 `Choice b.` (Distant objects are similar to nearby objects.)

4 `Choice d.` (100au)

5 `Choice a.` (10 parsecs.)

Where to find these questions in the website

1 Module 022.204

2 Module 022.402

3 Module 022.503

4 Module 022.102

5 Module 022.301

Module 023

Answers

1 `Choice d.` (K)

2 `Choice b.` (G0)

3 `Choice b.` (red.)

Where to find these questions in the website

1 Module 023.202

2 Module 023.301

3 Module 023.102

Module 024

Answers

1 `Choice d.` (D)

2 `Choice b.` (on the left side.)

3 `Choice c.` (hotter and brighter.)

4 `Choice d.` (a red main sequence star.)

5 `Choice d.` (stellar spectra to locate stars in the HR diagram.)

Where to find these questions in the website

1 Module 024.102

2 Module 024.204

3 Module 024.303

4 Module 024.404

5 Module 024.501

Module 025

Answers

1 `Choice b.` (stays at the same point until it runs out of fuel.)

2 `Choice b.` (proton.)

3 `Choice b.` (0.009amu)

4 `Choice b.` (temperature.)

5 `Choice d.` (form more slowly and burn slower.)

Where to find these questions in the website

1 Module 025.403

2 Module 025.101

3 Module 025.201

4 Module 025.302

5 Module 025.502

Module 026

Answers

1 `Choice c.` (a star becomes a white dwarf.)

2 `Choice b.` (the exhaustion of helium at its core.)

3 `Choice d.` (dropping temperature and constant brightness.)

4 `Choice e.` (lower left corner.)

5 `Choice b.` (helium at their centers.)

Where to find these questions in the website

1 Module 026.403

2 Module 026.301

3 Module 026.103

4 Module 026.408

5 Module 026.202

Module 027

Answers

1 `Choice b.` (absorb energy and limit the core temperature.)

2 `Choice a.` (during supernova explosions of very massive stars.)

3 `Choice d.` (happens when the electrons are eliminated.)

4 `Choice a.` (a spectrum with no hydrogen lines and a standard maximum brightness.)

5 `Choice b.` (10-15 million years.)

Where to find these questions in the website

1 Module 027.202

2 Module 027.405

3 Module 027.301

4 Module 027.503

5 Module 027.103

Module 028

Answers

1 `Choice c.` (of natural origin.)

2 `Choice a.` (similar in size to a neutron star.)

3 `Choice c.` (have been detected because infalling matter emits X-rays.)

4 `Choice b.` (event horizon.)

Where to find these questions in the website

1 Module 028.102

2 Module 028.303

3 Module 028.401

4 Module 028.302

Module 029

Answers

1 `Choice d.` (mass.)

2 `Choice e.` (30,000 light years.)

3 `Choice c.` (Orion Arm.)

4 `Choice a.` (is thickest.)

5 `Choice a.` (both infrared light and radio waves.)

6 `Choice a.` (smaller, nearby objects.)

7 `Choice c.` (more luminous and less common.)

Where to find these questions in the website

1 Module 029.508

2 Module 029.304

3 Module 029.406

4 Module 029.104

5 Module 029.503

6 Module 029.402

7 Module 029.204